MATLAB® Demystified

David McMahon

New York Chicago San Francisco Lisbon London
Madrid Mexico City Milan New Delhi San Juan
Seoul Singapore Sydney Toronto

The McGraw·Hill Companies

Cataloging-in-Publication Data is on file with the Library of Congress

McGraw-Hill books are available at special quantity discounts to use as premiums and sales promotions, or for use in corporate training programs. For more information, please write to the Director of Special Sales, Professional Publishing, McGraw-Hill, Two Penn Plaza, New York, NY 10121-2298. Or contact your local bookstore.

MATLAB® Demystified

567890 DOC DOC 0

ISBN-13: 978-0-07-148551-7
ISBN-10: 0-07-148551-1

Sponsoring Editor
Judy Bass

Editorial Supervisor
Janet Walden

Project Manager
Vasundhara Swahney, International Typesetting and Composition

Technical Editor
Rayjan Wilson

Copy Editor
Raina Trivedi

Proofreader
Ragini Pandey

Indexer
WordCo Indexing Services

Production Supervisor
George Anderson

Composition
International Typesetting and Composition

Illustration
Adam Brill

Art Director, Cover
Jeff Weeks

Cover Illustration
Lance Lekander

This book is dedicated to my parents, who were gracious enough to put up with my taking so long to find my way in life

ABOUT THE AUTHOR

David McMahon, Ph.D., is a physicist and researcher at Sandia National Laboratories. He is the author of *Linear Algebra Demystified, Quantum Mechanics Demystified, Relativity Demystified, Signals and Systems Demystified, and Statics and Dynamics Demystified.*

CONTENTS

PREFACE

MATLAB is one of the most widely used computational tools in science and engineering. No matter what your background—be it physics, chemistry, math, or engineering—it would behoove you to at least learn the basics of this powerful tool.

There are three good reasons to learn a computational mathematics tool. The first is that it serves as a background check for work you might be doing by hand. If you are a student, it's nice to have a back up that you can use to check your answers. I advise that you don't become co-dependent on a computational tool or trust it as though it were an Oracle. Do your work by hand when requested by your professors and just use MATLAB or any other tool to check your work to make sure it's correct.

The second reason is having a tool like MATLAB is priceless for generating plots and for doing numerical methods. Instead of having to go through a tedious process of plotting something by hand you can just have MATLAB generate any nice plot you desire.

Thirdly, the bottom line is that at some point in your career you will have to use a computational mathematics tool. If you're a professor doing theoretical work, at one point or another you are going to be working on a project where analytical solutions are not possible. If you work in industry or in a national lab, chances are the work you're doing can't be done by hand and will require a numerical solution. MATLAB is widely used in universities, in national laboratories and at private companies. Knowing MATLAB will definitely be a plus on your resume.

Now a word about this particular book. This book is aimed squarely at the MATLAB beginner. The purpose is not to wow experts with complicated solutions built with MATLAB. Rather, the purpose of this book is to introduce a person new to MATLAB to the world of computational mathematics. The approach taken here is to set about learning how to use MATLAB to do some basic things-plot functions, solve algebraic equations, compute integrals and solve differential equations for example. So the examples we present in this book are going to be simple and aimed at the novice. If you have never touched MATLAB before or are having lot's of trouble with it, this book will help you build a basic skill set that can be used to master it. The book is a stepping stone to mastery and nothing more.

ACKNOWLEDGMENTS

I would like to thank Rayjan Wilson for his thorough and thoughtful review of this manuscript. His insightful comments and detailed review were vital to making this manuscript a success.

CHAPTER 1

The MATLAB Environment

We begin our tour of MATLAB by considering the basic structure of the user interface. We'll learn how to enter commands, create files, and do other sorts of mundane tasks that we'll need to know before we can tackle solving mathematics problems. The elements covered in this chapter will be used throughout the book and indeed throughout the lifetime of your MATLAB use. In this book we are going to cover a core bit of knowledge about MATLAB to get you started using it. Our approach in this book is to take a few small bites in each chapter so that you can learn how to do a few important tasks at a time. At the end of the book you won't be a MATLAB expert, but you'll be on your way to getting comfortable with it and will know how to accomplish lots of common tasks, helping you make progress in your class at school or making it easier to grab a thick hard-to-read MATLAB book at the office so you can do real computing. Anyway, let's begin by considering the main MATLAB screen you see when you start the program.

Overview of the User Interface

In this book we will assume that you are using Windows, although that won't be relevant for the most part. Please note that we will be using MATLAB version 7.1 in this book. MATLAB is started just like any other Windows program. Just go to your program files menu and search for the MATLAB folder. When you click on it, you will see several options depending on your installation, but you will have at least the following three options

- MATLAB (version number)
- M-file editor
- Uninstaller

To start the program, you select MATLAB (7.1). The default MATLAB desktop will then open on your screen (see Figure 1-1). As shown in the figure, the screen is divided into three main elements. These are

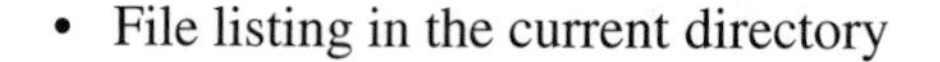

- File listing in the current directory

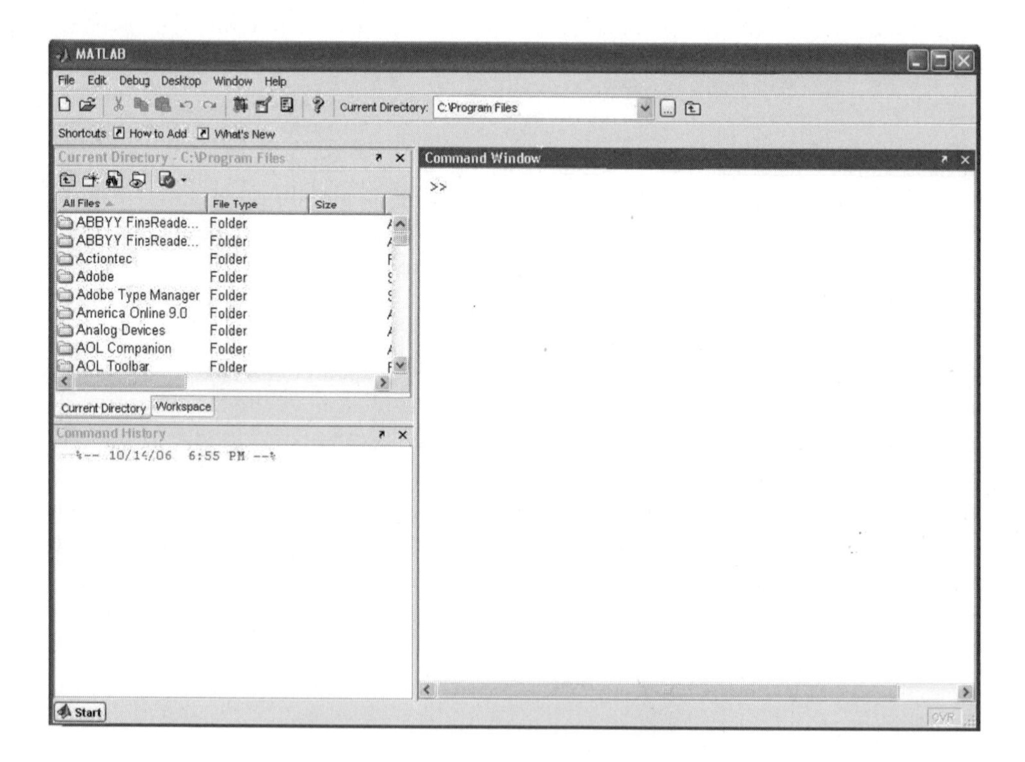

Figure 1-1 The MATLAB desktop

- Command History Window
- Command Window

The standard mix of menus appears on the top of the MATLAB desktop that allows you to do things like file management and debugging of files you create. You will also notice a drop-down list on the upper right side of the desktop that allows you to select a directory to work in. The most important item of business right now is the *Command Window*.

Command Window and Basic Arithmetic

The Command Window is found on the right-hand side of the MATLAB desktop. Commands are entered at the prompt with looks like two successive "greater than" signs:

```
>>
```

Let's start by entering a few really basic commands. If you want to find the value of a numerical expression, simply type it in. Let's say we want to know the value of 433.12 multiplied by 15.7. We type 433.12 * 15.7 at the MATLAB prompt and hit the ENTER key. The result looks like this:

```
>> 433.12*15.7
ans =
   6.8000e+003
```

MATLAB spits out the answer to our query conveniently named *ans*. This is a *variable* or symbolic name that can be used to represent the value later. Chances are we will wish to use our own variable names. So for example, we might want to call a variable x. Suppose we want to set it equal to five multiplied by six. To do this, we type the input as

```
>> x=5*6
x =
   30
```

Once a variable has been entered into the system, we can refer to it later. Suppose that we want to compute a new quantity that we'll call y, which is equal to x multiplied by 3.56. Then we type

```
>> y = x * 3.56
y =
   106.8000
```

Now, you will notice that in this example we put spaces in between each term in our equation. This was only done to enhance the readability and professional appearance of our output. MATLAB does not require you to include these spaces in your input. We could just as well type $y = x * 3.56$ as $y = x * 3.56$; however, the latter presentation is cleaner and easier to read. When your expressions get complicated, it will be more important to keep things neat so it's advisable to include the spaces.

Let's summarize basic arithmetical input in MATLAB. To write the multiplication *ab*, in MATLAB we type

```
a * b
```

For division, the quantity $\frac{a}{b}$ is typed as

```
a / b
```

This type of division is referred to as *right division.* MATLAB also allows another way to enter division, called *left division.* We can enter the quantity $\frac{b}{a}$ by typing the slash mark used for division in the opposite way, that is, we use a back slash instead of a forward slash

```
a \ b
```

Exponentiation a^b is entered in the following way

```
a ^ b
```

Finally, addition and subtraction are entered in the usual way

```
a + b
a - b
```

The precedence followed in mathematical operations by MATLAB is the same used in standard mathematics, but with the following caveat for left and right division. That is, exponentiation takes precedence over multiplication and division, which fall on equal footing. Right division takes precedence over left division. Finally, addition and subtraction have the lowest precedence in MATLAB. To override precedence, enclose expression in parentheses.

EXAMPLE 1-1

Use MATLAB to evaluate

$$5\left(\frac{3}{4}\right)+\frac{9}{5} \text{ and } 4^3\left[\frac{3}{4}+\frac{9}{(2)3}\right]$$

SOLUTION 1-1

The command required to find the value of the first expression is

```
>> 5*(3/4) + 9/5
ans =
   5.5500
```

For the second expression, we use some parentheses along with the exponentiation operator *a* ^ *b*. Although this is a simple expression, let's enter it in pieces to get used to using variables. We obtain

```
>> r = 4^3
r =
     64
>> s = 3/4 + 9/(2*3)
s =
     2.2500
>> t=r*s
t =
     144
```

The Assignment Operator

The equals sign "=" is known as the *assignment operator*. While it does what you think it does much of the time, that is, describes an equation, at other times in MATLAB it's more appropriate to think of it as an instruction to assign a value to a variable the way you would in a computer program. The distinction between the two interpretations can be illustrated in the following way. If you type

```
x + 6 = 90
```

in MATLAB, you get the following response

```
??? x+6=90
Error: The expression to the left of the equals sign is not
a valid target for an assignment.
```

So while the expression is a completely valid equation you could write down if doing algebra on paper, MATLAB doesn't know what to do with it. On the other hand, MATLAB is completely happy if you *assign* the value 90 – 6 to the variable *x* by writing

```
x = 90 - 6
```

Another way that the assignment operator works more like an assignment in a computer program is in a recursive type assignment to a variable. That is, MATLAB allows you to write

```
x = x + 4
```

if we have previously defined the variable *x*. For example, the following sequence is completely valid

```
>> x = 34^2
x =
        1156
>> x = x + 4
x =
        1160
```

To use a variable on the right-hand side of the assignment operator, we must assign a value to it beforehand. So while the following command sequence will generate an error

```
>> x = 2
x =
     2
>> t = x + a
??? Undefined function or variable 'a'.
```

The following sequence does not

```
>> x = 2
x =
     2
>> a = 3.5
a =
     3.5000
>> t = x + a
t =
     5.5000
```

In many instances, it is not desirable to have MATLAB spit out the result of an assignment. To suppress MATLAB output for an expression, simply add a semicolon (;) after the expression. In the following command sequence, first we just type in the assignment $x = 3$. MATLAB duly reports this back to us. On the next line, we enter $x = 3$; so that MATLAB does not waste space by telling us something we already know. Instead it comes back with the command prompt waiting for our next input:

```
>> x = 3
x =
     3
```

```
>> x = 3;
>>
```

We can include multiple assignments on the same line. For example, the following expressions are valid

```
>> x = 2; y = 4; z = x*y
z =
     8
```

Notice the two semicolons, they tell MATLAB we don't want to see the values of *x* and *y*.

When doing a lot of calculations, you may end up with a large number of variables. You can refresh your memory by typing *who* in the MATLAB command window. Doing this will tell MATLAB to display all of the variable names you have used up to this point. For instance, in our case we have

```
>> who
Your variables are:
V  a     ans   r     s     t     x     y     z
```

By typing *whos,* we get a bit more information. This will tell us the variables currently in memory, their type, how much memory is allocated to each variable, and whether or not they are *complex* (see below). In our case we have

```
>> whos
Name       Size             Bytes   Class

V          1x1                  8   double array
a          1x1                  8   double array
ans        1x1                 16   double array (complex)
r          1x1                  8   double array
s          1x1                  8   double array
t          1x1                  8   double array
x          1x1                  8   double array
y          1x1                  8   double array
z          1x1                  8   double array

Grand total is 9 elements using 80 bytes
```

Now suppose we want to start all over. We can do this by issuing a *clear* command. Clear can be applied globally by simply typing *clear* and then hitting the ENTER key, or to specific variables by typing clear followed by a space delimited variable list. If we wanted to reset or clear the variables *x, y,* and *z* that we have been using, then we could type

```
clear x y z
```

MATLAB will simply return the command prompt and won't say anything else, but if you try to use these variables again without assigning them values it will be as if they had not been seen before.

Long assignments can be extended to another line by typing an *ellipsis* which is just three periods in a row. For example

```
>> FirstClassHolders = 72;
>> Coach = 121;
>> Crew = 8;
>> TotalPeopleOnPlane = FirstClassHolders + Coach...
+ Crew

TotalPeopleOnPlane =

   201
```

The ellipsis follows *Coach* on the line used to define *TotalPeopleOnPlane.* After you type the ellipsis, just hit the ENTER key. MATLAB will move to the following line awaiting further input.

OK, something you are probably wondering, as I was when I started using MATLAB, was how in the world do you control the way numbers are displayed on the screen? So far in our examples, MATLAB has been spitting out numbers with four decimal places. This is known as *short* format in MATLAB. It's the default in MATLAB and if that's all the precision you require, then you don't have to do anything. If you want more, then you can tell MATLAB to add more digits to the right of the decimal point by using the *format* command. If we want 16 digits instead of 4, we type *format long.* To see how this works, look at the following calculation, displayed in both formats

```
>> format long
>> x = 3 + 11/16 + 2^1.2
x =
   5.98489670999407

>> format short
>> x = 3 + 11/16 + 2^1.2
x =
   5.9849
```

Comparing the long and short formats, notice that the fourth decimal place was rounded up to nine when *format short* was used. If you want to do financial calculations, you can use the *format bank* command. As expected, this rounds everything off to two decimal places.

```
>> format bank
>> hourly = 35.55
```

```
hourly =
         35.55

>> weekly = hourly*40
weekly =
         1422.00
```

MATLAB displays large numbers using exponential notation. That is it represents 5.4387×10^3 as $5.4387e + 003$. If you want all numbers to be represented in this fashion, you can do so. This type of notation can also be defined using the short or long formats. For short (four decimal places plus the exponent) you can type *format short e*. To allow 15 decimal digits plus the exponent, type *format long e*. Here is an example of the short exponent format

```
>> format short e
>> 7.2*3.1
ans =
   2.2320e+001
```

If you type *format rat,* then MATLAB will find the closest rational expression it can that corresponds to the result of a calculation. Pretty neat tool huh? Let's repeat the previous calculation

```
>> format rat
>> 7.2*3.1
ans =
     558/25
```

Basic Mathematical Definitions

MATLAB comes with many basic or familiar mathematical quantities and functions built in. Let's show how to use π in an example.

EXAMPLE 1-2
Find the volume of a sphere of radius 2 m.

SOLUTION 1-2
The volume of a sphere is given by

$$V = \frac{4}{3}\pi R^3$$

Of course MATLAB comes with π predefined. To use it, we just type *pi*. So after defining a variable to hold the radius, we can find the volume by typing

```
>> r = 2;
>> V = (4/3) *pi*r^3
V =
   33.5103
```

Another famous number that shows up in many mathematical applications is the exponential function. That is, $e \approx 2.718$. We can reference e in MATLAB by typing *exp*(a) which gives us the value of e^a. Here are a few quick examples

```
>> exp(1)
ans =
     2.7183

>> exp(2)
ans =
     7.3891
```

To find the square root of a number, we type *sqrt*. For example

```
>> x = sqrt(9)
x =
     3

>> y = sqrt(11)
y =
     3.3166
```

To find the *natural log* of a number x, type *log*(x).

```
>> log(3.2)
ans =
     1.1632
>> x = 5; log(5)
ans =
     1.6094
```

If you want the base ten logarithm, type *log*10(x)

```
>> x = 3; log10(x)
ans =
     0.4771
```

MATLAB comes equipped with the basic trig functions and their inverses, taking radian argument by default. These are typed in lower case using the standard notation. For instance

```
>> cos(pi/4)
ans =
    0.7071
```

To use an inverse of a trig function, add on an *a* before the name of the trig function. For example, to compute the inverse tangent of a number we can use the following

```
>> format rat
>> atan(pi/3)
ans =
    1110/1373
```

Complex Numbers

We can also enter complex numbers in MATLAB. To remind members of our audience who are Aggie graduates, the square root of –1 is defined as

$$i = \sqrt{-1}$$

A complex number is one that can be written in the form $z = x + iy$, where x is the real part of z and y is the imaginary part of z. It is easy to enter complex numbers in MATLAB, by default it recognizes i as the square root of minus one. We can do calculations with complex numbers in MATLAB pretty easily. For example

```
a = 2 +3i
b = 1 – i
⇒ a + b = 3 + 2i
```

Let's verify this in MATLAB. It is not necessary to add spaces or include a multiplication symbol (*) when typing i in MATLAB.

```
>> format short
>> a = 2 + 3i;
>> b = 1 - i;
>> c = a + b
c =
   3.0000 + 2.0000i
```

Fixing Typos

It's going to be a fact that now and then you are going to type in an expression with an error. If you hit the ENTER key and then later realize what happened, it's not necessary to retype the line. Just use your arrow keys to move back up to the offending line. Fix your error, and then hit ENTER again and MATLAB will correct the output.

Some File Basics

Let's round out the chapter by considering some basic operations with files. MATLAB wouldn't be *that* useful if you couldn't save and retrieve your work right? Let's say you want to save all the expressions and variables you have entered in the command window for use at a later time. You can do this by executing the following actions:

1. Click on the File pull-down menu
2. Select Save Workspace As…
3. Type in a file name
4. Click on the Save button

This method creates a MATLAB file which has a .MAT file extension in Windows. If you save a file this way, you can retrieve all the commands in it and work with it again just like you can when working with files in any other computer program.

Sometimes, especially when working on complicated projects, you won't want to sit there and type every expression in a command window. It might be more appropriate to type a long sequence of operations and store them in a file that can be executed with a single command in the command window. This is done by creating a *script file*. This type of file is known as a MATLAB program and is saved in a file format with a .M extension. For this reason, we also call them M-files. We can also create M-files that are *function* files.

From what we've done so far, you already know how to create a script file. All a script file comes down to is a saved sequence of MATLAB commands. Let's create a simple script file that will compute e^x for a few values of x. First, open the MATLAB editor. Either

- Click New → M-File under the File pull-down menu
- Or click on the New File icon on our toolbar at the top of the screen

Now type in the following lines:

```
% script file example1.m to compute exponential of a set of numbers
x = [1:2:3:4];
y = exp(x)
```

Notice the first line begins with a % sign. This line is a *comment*. This is a line of text that is there for our benefit, it's a descriptive note that MATLAB ignores. The next line creates an *array* or set of numbers. An array is denoted using square braces [] and by delimiting the elements of the array with colons or commas. The final line will tell MATLAB to calculate the exponential of each member of the array, in other words the values e^1, e^2, e^3, e^4. Save the file by clicking the Save icon in the file editor or by selecting Save As from the File pull-down menu. Save the file as *example1.m* in your MATLAB directory.

Now return to the MATLAB desktop command window. Type in *example1*. If you did everything right, then you will see the following output

```
>> example1
y =
   2.7183        7.3891        20.0855       54.5982
```

We can also use M-files to create and store data. Following an example from another McGraw-Hill book on MATLAB, let's create a set of temperatures that we will store in a file. We do this by creating a list of temperatures in the file editor

```
temps = [32,50,65,70,85]
```

Now we save this as a file that we'll call *TemperatureData.m.* We store this file in the MATLAB directory. To access it in the command window, we just type the name of the file. MATLAB responds by spitting out the list of numbers:

```
>> TemperatureData
temps =
       32         50         65         70         85
```

Now we can use the data by referring to the array name used in the file. Let's create another set of numbers called Celsius that converts these familiar temps into the European style Celsius temperatures we are so familiar with. This can be done with the following command

```
>> CelsiusTemps = (5/9) * (temps - 32)
CelsiusTemps =
   0      10.0000      18.3333      21.1111      29.4444
```

Later, we will investigate programming in MATLAB and we will show you how to create functions that can be called later in the command window.

Ending Your MATLAB Session

OK we have gotten started with a few very basic MATLAB commands. You might want to save your work and then shut down MATLAB. How do you get it off your screen? Well you can end your MATLAB session by selecting *exit* from the File pull-down menu, just like you would with any other program. Optionally, you can type *quit* in the command window and MATLAB will close.

Quiz

Use MATLAB to calculate the following quantities:

1. $5\frac{11}{14}$
2. $5\frac{8}{3}+3^7$
3. $9^{1.25}$
4. True or False. If y has not been assigned a value, MATLAB will allow you to define the equation $x = y$ ^2 to store in memory for later use.
5. If the volume of a cylinder of height h and radius r is given by $V = \pi r^2 h$, use MATLAB to find the volume enclosed by a cylinder that is 12 cm high with a diameter of 4 cm.
6. Use MATLAB to compute the sin of $\pi/3$ expressed as a rational number.
7. Create a MATLAB m file to display the results of $\sin(\pi/4)$, $\sin(\pi/3)$, $\sin(\pi/2)$ as rational numbers.

CHAPTER 2

Vectors and Matrices

One area where MATLAB is particularly useful is in the computer implementation of linear algebra problems. This is because MATLAB has exceptional capabilities for handling arrays of numbers, making it a useful tool for many scientific and engineering applications.

Vectors

A *vector* is a one-dimensional array of numbers. MATLAB allows you to create *column vectors* or *row vectors*. A column vector can be created in MATLAB by enclosing a set of semicolon delimited numbers in square brackets. Vectors can have any number of elements. For example, to create a column vector with three elements we write:

```
>> a = [2; 1; 4]
a =
     2
     1
     4
```

Basic operations on column vectors can be executed by referencing the variable name used to create them. If we multiply a column vector by a number, this is called *scalar multiplication.* Suppose that we wanted to create a new vector such that its components were three times the components in the vector **a** we just created above. We could start by defining a scalar variable (remember that a semicolon after a command suppresses the output):

```
>> c = 3;
```

Next, we just perform the operation treating *a like another variable*:

```
>> b = c*a
b =
     6
     3
    12
```

To create a row vector, we enclose a set of numbers in square brackets but this time use a space or comma to delimit the numbers. For example:

```
>> v = [2 0 4]
v =
     2     0     4
```

Or using commas:

```
>> w = [1,1,9]
w =
     1     1     9
```

Column vectors can be turned into row vectors and vice versa using the *transpose* operation. Suppose that we have a column vector with *n* elements denoted by:

$$v = \begin{bmatrix} v_1 \\ v_2 \\ \vdots \\ v_n \end{bmatrix}$$

Then the transpose is given by:

$$v^T = \begin{bmatrix} v_1 & v_2 & \cdots & v_n \end{bmatrix}$$

In MATLAB, we represent the transpose operation with a single quote or tickmark (‘). Taking the transpose of a column vector produces a row vector:

```
>> a = [2; 1; 4];
>> y = a'
y =
     2     1     4
```

Now let’s take the transpose of a row vector to produce a column vector:

```
>> Q = [2 1 3]
Q =
     2     1     3
>> R = Q'
R =
     2
     1
     3
```

It is also possible to add or subtract two vectors to produce another. In order to perform this operation the vectors must both be of the same type and the same length, so we can add two column vectors together to produce a new column vector or we can add two row vectors to produce a new row vector, for example. This can be done referencing the variables only, it is not necessary for the user to list the components. For example, let’s add two column vectors together:

```
>> A = [1; 4; 5];
>> B = [2; 3; 3];
>> C = A + B
C =
     3
     7
     8
```

Now let’s subtract one row vector from another:

```
>> W = [3,0,3];
>> X = [2,1,1];
>> Y = W - X
Y =
     1    -1     2
```

Creating Larger Vectors from Existing Variables

MATLAB allows you to append vectors together to create new ones. Let **u** and **v** be two column vectors with m and n elements respectively that we have created in MATLAB. We can create a third vector **w** whose first m elements are the elements of **u** and whose next n elements are the elements of **v**. The newly created column vector has $m + n$ elements. This is done by writing w = [u; v]. For example:

```
>> A = [1; 4; 5];
>> B = [2; 3; 3];
>> D = [A;B]
D =
     1
     4
     5
     2
     3
     3
```

This can also be done using row vectors. To create a row vector **u** with $m + n$ elements from a vector **r** with m elements and a vector **s** with n elements, we write u = [r, s]. For example:

```
>> R = [12, 11, 9]
>> S = [1, 4];
>> T = [R, S]
T =
    12    11     9     1     4
```

Creating Vectors with Uniformly Spaced Elements

It is possible to create a vector with elements that are uniformly spaced by an increment q, where q is any real number. To create a vector **x** with uniformly spaced elements where x_i is the first element and x_e is the final element, the syntax is:

$$x = [x_i : q : x_e]$$

For example, we can create a list of even numbers from 0 to 10 by writing:

```
>> x = [0:2:10]
x =
     0     2     4     6     8    10
```

We have already been using this technique in the last chapter to create a list of values for plotting purposes. Let's see how this works behind the scenes by looking at some vectors with a small number of elements. First we create a set of x values:

```
>> x = [0:0.1:1]
x =
```

Columns 1 through 10

```
         0    0.1000    0.2000    0.3000    0.4000    0.5000
0.6000    0.7000    0.8000    0.9000
```

Column 11

```
    1.0000
```

The set of x values can be used to create a list of points representing the values of some given function. For example, suppose that $y = e_x$. Then we have:

```
>> y = exp(x)
y =
```

Columns 1 through 10

```
    1.0000    1.1052    1.2214    1.3499    1.4918    1.6487
1.8221    2.0138    2.2255          2.4596
```

Column 11

```
    2.7183
```

Or we could have $y = x^2$:

```
>> y = x^2
y =
```

Columns 1 through 10

```
         0    0.0100    0.0400    0.0900    0.1600    0.2500
0.3600    0.4900    0.6400    0.8100
```

Column 11

```
    1.0000
```

An aside—note that when squaring a vector in MATLAB, a period must precede the power operator (^). If we just enter >> y = x^2, MATLAB gives us an error message:

```
??? Error using ==> mpower
Matrix must be square.
```

Returning to the process of creating an array of uniformly spaced elements, be aware that you can also use a negative increment. For instance, let's create a list of numbers from 100 to 80 decreasing by 5:

```
>> u = [100:-5:80]
u =
   100     95     90     85     80
```

When looking at plotting we have also seen how to create a row vector with elements from a to b with n regularly spaced elements using the *linspace* command. To review, linspace(a,b) creates a row vector of 100 regularly spaced elements between a and b, while linspace(a,b,n) creates a row vector of n regularly spaced elements between a and b. In both cases MATLAB determines the increment in order to have the correct number of elements.

MATLAB also allows you to create a row vector of n logarithmically spaced elements by typing

```
logspace(a,b,n)
```

This creates n regularly spaced elements between 10^a and 10^b. For example:

```
>> logspace(1,2,5)
ans =
   10.0000   17.7828   31.6228   56.2341  100.0000
```

Or another example:

```
>> logspace(-1,1,6)
ans =
    0.1000    0.2512    0.6310    1.5849    3.9811   10.0000
```

Characterizing a Vector

The *length* command returns the number of elements that a vector contains. For example:

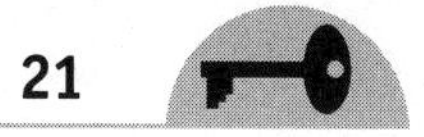

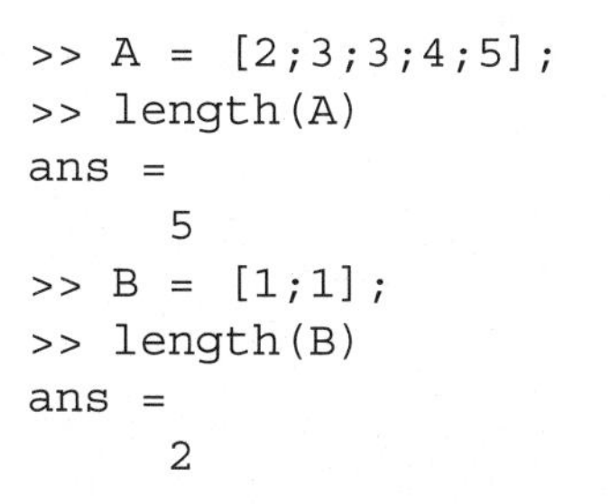

```
>> A = [2;3;3;4;5];
>> length(A)
ans =
     5
>> B = [1;1];
>> length(B)
ans =
     2
```

The length command can be applied to row and column vectors (see “Basic Operations with Matrices”, later in this chapter) and, as we will see below, to matrices.

We can find the largest and smallest elements in a vector using the *max* and *min* commands. For example:

```
>> A = [8 4 4 1 7 11 2 0];
>> max(A)
ans =
    11
>> min(A)
ans =
     0
```

To find the magnitude of a vector we can employ two operations. Recall that the magnitude of a vector

$$v = \begin{bmatrix} v_1 \\ v_2 \\ \vdots \\ v_n \end{bmatrix}$$

is given by:

$$|v| = \sqrt{v_1^2 + v_2^2 + \cdots v_n^2}$$

To perform this operation, we will first take the dot product of a vector with itself. This is done by using array multiplication (.*). First let’s define a vector:

```
>> J = [0; 3; 4];
```

Now we can do array multiplication:

```
>> J.*J
ans =
```

```
    0
    9
   16
```

This produces a vector whose elements are $v_1^2, v_2^2, \ldots$. To get the summation we need, we can use the sum operator:

```
>> a = sum(J.*J)
a =
    25
```

Then the magnitude of the vector is the square root of this quantity:

```
>> mag = sqrt(a)
mag =
     5
```

If a vector contains complex numbers, then more care must be taken when computing the magnitude. When computing the row vector, we must compute the complex conjugate transpose of the original vector. For example, if:

$$u = \begin{bmatrix} i \\ 1+2i \\ 4 \end{bmatrix}$$

Then to compute the magnitude, we need the following vector:

$$u^\dagger = [-i \quad 1-2i \quad 4]$$

Then the summation we need to compute is:

$$u^\dagger u = [-i \quad 1-2i \quad 4]\begin{bmatrix} i \\ 1+2i \\ 4 \end{bmatrix} = (-i)(i) + (1-2i)(1+2i) + (4)(4) = 22$$

Hence the magnitude of a vector with complex elements is:

$$|u| = \sqrt{u^\dagger u} = \sqrt{22}$$

You can see how using the complex conjugate transpose when computing our sum ensures that the magnitude of the vector will be real. Now let's see how to do this in MATLAB. First let's enter this column vector:

```
>> u = [i; 1+2i; 4];
```

If we just compute the sum of the vector multiplication as we did in the previous example, we get a complex number that won't work:

```
>> sum(u.*u)
ans =
  12.
```

So let's define another vector which is the complex conjugate transpose of **u**. MATLAB does this automatically with the transpose operator:

```
>> v = u'
v =
        0 - 1.0000i   1.0000 - 2.0000i   4.0000
```

Now we can perform our sum:

```
>> b = sum(v.*u)
??? Error using ==> times
```

Matrix dimensions must agree.

Unfortunately it looks like we've been led down a blind alley! It appears there isn't quite a one to one correspondence to what we would do on paper. How can we get around this? Let's just compute the complex conjugate of the vector, and form the sum. We can get the complex conjugate of a vector with the *conj* command:

```
>> v = conj(u)
v =
        0 - 1.0000i
   1.0000 - 2.0000i
   4.0000
```

Now we obtain the correct answer, and can get the magnitude:

```
>> b = sum(v.*u)
b =
    22
>> magu = sqrt(b)
magu =
    4.6904
```

Of course this could all be done in one step by writing:

```
>> c = sqrt(sum(conj(u).*u))
c =
    4.6904
```

Here we are actually doing things the hard way—just to illustrate the method and some MATLAB commands. In the next section we will see how to compute the magnitude of a vector automatically.

We can use the *abs* command to return the absolute value of a vector, which is a vector whose elements are the absolute values of the elements in the original vector, i.e.:

```
>> A = [-2 0 -1 9] >> B = abs(A)
B =
     2     0     1     9
```

Vector Dot and Cross Products

The *dot product* between two vectors $A = (a_1\ a_2\ \ldots\ a_n)$ and $B = (b_1\ b_2\ \ldots\ b_n)$ is given by

$$\boldsymbol{A} \cdot \boldsymbol{B} = \sum_i \boldsymbol{a}_i \boldsymbol{b}_i$$

In MATLAB, the dot product of two vectors **a, b** can be calculated using the dot(a,b) command.

The dot product between two vectors is a scalar, i.e. it's just a number. Let's compute a simple example using MATLAB:

```
>> a = [1;4;7]; b = [2;-1;5];
>> c = dot(a,b)
c =
    33
```

The dot product can be used to calculate the magnitude of a vector. All that needs to be done is to pass the same vector to both arguments. Consider the vector in the last section:

```
>> J = [0; 3; 4];
```

Calling dot we obtain:

```
>> dot(J,J)
ans =
    25
```

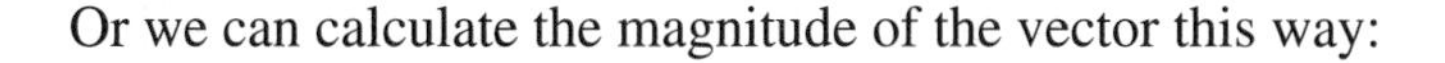

Or we can calculate the magnitude of the vector this way:

```
>> mag = sqrt(dot(J,J))
mag =
     5
```

The dot operation works correctly with vectors that contain complex elements:

```
>> u = [-i; 1 + i; 4 + 4*i];
>> dot(u,u)
ans =
    35
```

Another important operation involving vectors is the *cross product.* To compute the *cross product*, the vectors must be three dimensional. For example:

```
>> A = [1 2 3]; B = [2 3 4];
>> C = cross(A,B)
C =
    -1     2    -1
```

Referencing Vector Components

MATLAB has several techniques that can be used to reference one or more of the components of a vector. The *i*th component of a vector **v** can be referenced by writing **v(i)**. For example:

```
>> A = [12; 17; -2; 0; 4; 4; 11; 19; 27];
>> A(2)
ans =
    17
>> A(8)
ans =
    19
```

Referencing the vector with a colon, such as v(:); tells MATLAB to list all of the components of the vector:

```
>> A(:)
ans =
    12
    17
```

```
   -2
    0
    4
    4
   11
   19
   27
```

We can also pick out a range of elements out of a vector. The example we've been working with so far in this section is a vector **A** with nine components. We can reference components four to six by writing A(4:6) and use these to create a new vector with three components:

```
>> v = A(4:6)
v =
     0
     4
     4
```

In the next section, we will see how these techniques can be used to reference the columns or rows in an array of numbers called a matrix.

Basic Operations with Matrices

A *matrix* is a two-dimensional array of numbers. To create a matrix in MATLAB, we enter each row as a sequence of comma or space delimited numbers, and then use semicolons to mark the end of each row. For example, consider:

$$A = \begin{bmatrix} -1 & 6 \\ 7 & 11 \end{bmatrix}$$

This matrix is entered in MATLAB using the following syntax:

```
>> A = [-1,6; 7, 11];
```

Or consider the matrix:

$$B = \begin{bmatrix} 2 & 0 & 1 \\ -1 & 7 & 4 \\ 3 & 0 & 1 \end{bmatrix}$$

We enter it in MATLAB in the following way:

```
>> B = [2,0,1;-1,7,4; 3,0,1]
```

Many of the operations that we have been using on vectors can be extended for use with matrices. After all a column vector with n elements is just a matrix with one column and n rows while a row vector with n elements is just a matrix with one row and n columns. For example, scalar multiplication can be carried out referencing the name of the matrix:

```
>> A = [-2 2; 4 1]
A =
    -2     2
     4     1
>> C = 2*A
C =
    -4     4
     8     2
```

If two matrices have the same number of rows and columns, we can add and subtract them:

```
>> A = [5 1; 0 9];
>> B = [2 -2; 1 1];
>> A + B
ans =
     7    -1
     1    10
>> A - B
ans =
     3     3
    -1     8
```

We can also compute the transpose of a matrix. The transpose operation switches the rows and columns in a matrix, for example:

$$A = \begin{bmatrix} -1 & 2 & 4 \\ 0 & 1 & 6 \\ 2 & 7 & 1 \end{bmatrix}, \Rightarrow A^T = \begin{bmatrix} -1 & 0 & 2 \\ 2 & 1 & 7 \\ 4 & 6 & 1 \end{bmatrix}$$

We can take the transpose of a matrix using the same notation used to calculate the transpose of a vector:

```
>> A = [-1 2 0; 6 4 1]
A =
    -1     2     0
     6     4     1
>> B = A'
B =
    -1     6
     2     4
     0     1
```

If the matrix contains complex elements, the transpose operation will compute the conjugates:

```
>> C = [1+i, 4-i; 5+2*i, 3-3*i]
C =
   1.0000 + 1.0000i   4.0000 - 1.0000i
   5.0000 + 2.0000i   3.0000 - 3.0000i
>> D = C'
D =
   1.0000 - 1.0000i   5.0000 - 2.0000i
   4.0000 + 1.0000i   3.0000 + 3.0000i
```

If we want to compute the transpose of a matrix with complex elements without computing the conjugate, we use (.'):

```
>> D = C'
D =
   1.0000 + 1.0000i   5.0000 + 2.0000i
   4.0000 - 1.0000i   3.0000 - 3.0000i
```

We can perform *array* multiplication. It is important to recognize that this is *not* matrix multiplication. We use the same notation used when multiplying two vectors together (.*). For example:

```
>> A = [12 3; -1 6]; B = [4 2; 9 1];
>> C = A.*B
C =
    48     6
    -9     6
```

As you can see, what the operation $A.*B$ does is it performs component by component multiplication. Abstractly it works as follows:

$$A.*B = \begin{pmatrix} a_{11} & a_{12} \\ a_{21} & a_{22} \end{pmatrix} \begin{pmatrix} b_{11} & b_{12} \\ b_{21} & b_{22} \end{pmatrix} = \begin{pmatrix} (a_{11})(b_{11}) & (a_{12})(b_{12}) \\ (a_{21})(b_{21}) & (a_{22})(b_{22}) \end{pmatrix}$$

We see how to perform matrix multiplication in the next section.

Matrix Multiplication

Consider two matrices A and B. If A is an $m \times p$ matrix and B is a $p \times n$ matrix, they can be multiplied together to produce an $m \times n$ matrix. To do this in MATLAB, we leave out the period (.) and simply write A*B. Keep in mind that if the dimensions of the two matrices are not correct, the operation will generate an error.

Let's consider two matrices:

$$A = \begin{pmatrix} 2 & 1 \\ 1 & 2 \end{pmatrix}, \quad B = \begin{pmatrix} 3 & 4 \\ 5 & 6 \end{pmatrix}$$

These are both 2×2 matrices, so matrix multiplication is permissible. First, let's do array multiplication so we can see the difference:

```
>> A = [2 1; 1 2]; B = [3 4; 5 6];
>> A.*B
ans =
      6     4
      5    12
```

Now we leave out the '.' character and execute matrix multiplication, which produces quite a different answer:

```
>> A*B
ans =
     11    14
     13    16
```

Here is another example. Consider:

$$A = \begin{pmatrix} 1 & 4 \\ 8 & 0 \\ -1 & 3 \end{pmatrix}, \quad B = \begin{pmatrix} -1 & 7 & 4 \\ 2 & 1 & -2 \end{pmatrix}$$

The matrix A is a 3×2 matrix, while B is a 2×3 matrix. Since the number of columns of A matches the number of rows of B, we can calculate the product AB. In MATLAB:

```
>> A = [1 4; 8 0; -1 3]; B = [-1 7 4; 2 1 -2];
>> C = A*B
C =
     7    11    -4
    -8    56    32
     7    -4   -10
```

While matrix multiplication is possible in this case, array multiplication is not. To use array multiplication, both row and column dimensions must agree for each array. Here is what MATLAB tells us:

```
>> A.*B
??? Error using ==> times
Matrix dimensions must agree.
```

More Basic Operations

MATLAB also allows several operations on matrices that you might not be immediately used to from your linear algebra background. For example, MATLAB allows you to add a scalar to an array (vector or matrix). This operation works by adding the value of the scalar to each component of the array. Here is how it works with a row vector:

```
>> A = [1 2 3 4];
>> b = 2;
>> C = b + A
C =
     3     4     5     6
```

We can also perform left and right division on an array. This works by matching component by component, so the arrays have to be of the same size. For example, we tell MATLAB to perform array right division by typing (./):

```
A = [2 4 6 8]; B = [2 2 3 1];
>> C = A./B
C =
     1     2     2     8
```

Array left division is indicated by writing C = A.\B (this is the same as C = B./A):

```
>> C = A.\B
C =
    1.0000    0.5000    0.5000    0.1250
```

Basically any mathematical operation you can think of can be implemented in MATLAB with arrays. For instance, we can square each of the elements:

```
>> B = [2 4; -1 6]
B =
     2     4
    -1     6
>> B.^2
ans =
     4    16
     1    36
```

Special Matrix Types

The *identity* matrix is a square matrix that has ones along the diagonal and zeros elsewhere. To create an $n \times n$ identity matrix, type the following MATLAB command:

```
eye(n)
```

Let's create a 4×4 identity matrix:

```
>> eye(4)
ans =
     1     0     0     0
     0     1     0     0
     0     0     1     0
     0     0     0     1
```

To create an $n \times n$ matrix of zeros, we type zeros(n). We can also create a $m \times n$ matrix of zeros by typing zeros(m,n). It is also possible to generate a matrix completely filled with 1's. Surprisingly, this is done by typing ones(n) or ones(m,n) to create $n \times n$ and $m \times n$ matrices filled with 1's, respectively.

Referencing Matrix Elements

Individual elements and columns in a matrix can be referenced using MATLAB. Consider the matrix:

```
>> A = [1 2 3; 4 5 6; 7 8 9]
A =
     1     2     3
     4     5     6
     7     8     9
```

We can pick out the element at row position m and column position n by typing A(m,n). For example:

```
>> A(2,3)
ans =
     6
```

To reference all the elements in the ith column we write A(:,i). For example, we can pick out the second column of A:

```
>> A(:,2)
ans =
     2
     5
     8
```

To pick out the elements in the ith through jth columns we type A(:,i:j). Here we return the second and third columns:

```
>> A(:,2:3)
ans =
     2     3
     5     6
     8     9
```

We can pick out pieces or submatrices as well. Continuing with the same matrix, to pick out the elements in the second and third rows that are also in the first and second columns, we write:

```
>> A(2:3,1:2)
ans =
     4     5
     7     8
```

We can change the value of matrix elements using these references as well. Let's change the element in row 1 and column 1 to –8:

```
>> A(1,1)  = -8
A =
    -8     2     3
     4     5     6
     7     8     9
```

To create an empty array in MATLAB, simply type an empty set of square braces []. This can be used to delete a row or column in a matrix. Let's delete the second row of A:

```
>> A(2,:)=[]
A =
    -8     2     3
     7     8     9
```

This has turned the formerly 3×3 matrix into a 2×3 matrix.

It's also possible to reference rows and columns in a matrix and use them to create new matrices. In this example, we copy the first row of A four times to create a new matrix:

```
>> E = A([1,1,1,1],:)
E =
    -8     2     3
    -8     2     3
    -8     2     3
    -8     2     3
```

This example creates a matrix out of both rows of A:

```
>> F = A([1,2,1],:)
F =
    -8     2     3
     7     8     9
    -8     2     3
```

Finding Determinants and Solving Linear Systems

The *determinant* of a square matrix is a number. For a 2×2 matrix, the determinant is given by:

$$D = \begin{vmatrix} a_{11} & a_{12} \\ a_{21} & a_{22} \end{vmatrix} = a_{11}a_{22} - a_{12}a_{21}$$

To calculate the determinant of a matrix A in MATLAB, simply write det(A). Here is the determinant of a 2×2 matrix:

```
>> A = [1 3; 4 5];
>> det(A)
ans =
     -7
```

In this example, we find the determinant of a 4×4 matrix:

```
>> B = [3 -1 2 4; 0 2 1 8; -9 17 11 3; 1 2 3 -3];
>> det(B)
ans =
   -533
```

Needless to say, such a calculation would be extremely tedious by hand. Determinants can be used to find out if a linear system of equations has a solution. Consider the following set of equations:

$$\begin{aligned} 5x + 2y - 9z &= 44 \\ -9x - 2y + 2z &= 11 \\ 6x + 7y + 3z &= 44 \end{aligned}$$

To find a solution to a system of equations like this, we can use two steps. First we find the determinant of the coefficient matrix A, which in this case is:

$$A = \begin{pmatrix} 5 & 2 & -9 \\ -9 & -3 & 2 \\ 6 & 7 & 3 \end{pmatrix}$$

The determinant is:

```
>> A = [5 2 -9; -9 -3 2; 6 7 3]
>> det(A)
ans =
   368
```

When the determinant is nonzero, a solution exists. This solution is the column vector:

$$\mathbf{x} = \begin{pmatrix} x \\ y \\ z \end{pmatrix}$$

MATLAB allows us to generate the solution readily using left division. First we create a column vector of the numbers on the right-hand side of the system. We find:

```
>> b = [44;11;5];
>> A\b
ans =
   -5.1250
    7.6902
   -6.0272
```

Finding the Rank of a Matrix

The *rank* of a matrix is a measure of the number of *linearly independent* rows or columns in the matrix. If a vector is linearly independent of a set of other vectors that means it cannot be written as a linear combination of them. Simple example:

$$u = \begin{pmatrix} 1 \\ -2 \end{pmatrix}, v = \begin{pmatrix} 3 \\ -4 \end{pmatrix}, w = \begin{pmatrix} 5 \\ -6 \end{pmatrix}$$

Looking at these column vectors we see that:

$$2u + v = w$$

Hence w is linearly dependent on u and v, since it can be written as a linear combination of them. On the other hand:

$$u = \begin{pmatrix} 2 \\ 0 \\ 0 \end{pmatrix}, v = \begin{pmatrix} 0 \\ -1 \\ 0 \end{pmatrix}, w = \begin{pmatrix} 0 \\ 0 \\ 7 \end{pmatrix}$$

form a linearly independent set, since none of these vectors can be written as a linear combination of the other two.

Consider the matrix:

$$A = \begin{pmatrix} 0 & 1 & 0 & 2 \\ 0 & 2 & 0 & 4 \end{pmatrix}$$

The second row of the matrix is clearly twice the first row of the matrix. Hence there is only one unique row and the rank of the matrix is 1. Let's check this in MATLAB. We compute the rank in the following way:

```
>> A = [0 1 0 2; 0 2 0 4];
>> rank(A)
ans =
     1
```

Another example:

$$B = \begin{pmatrix} 1 & 2 & 3 \\ 3 & 0 & 9 \\ -1 & 2 & -3 \end{pmatrix}$$

The third column is three times the first column:

$$\begin{matrix} 3 \\ 9 \\ -3 \end{matrix} = 3\begin{pmatrix} 1 \\ 3 \\ -1 \end{pmatrix}$$

Therefore, it's linearly dependent on the other two columns (add zero times the second column). The other two columns are linearly independent since there is no constant α such that:

$$\begin{pmatrix} 2 \\ 0 \\ 2 \end{pmatrix} = \alpha \begin{pmatrix} 1 \\ 3 \\ -1 \end{pmatrix}$$

So we conclude that there are two linearly independent columns, and rank(B) = 2. Let's check it in MATLAB:

```
>> B = [1 2 3; 3 0 9; -1 2 -3];
>> rank(B)
ans =
     2
```

Now let's consider the linear system of equations with m equations and n unknowns:

$$\mathbf{Ax} = \mathbf{b}$$

The *augmented* matrix is formed by concatenating the vector **b** onto the matrix **A**:

$$[\mathbf{A}\ \ \mathbf{b}]$$

The system has a solution if and only if rank(**A**) = rank(**A b**). If the rank is equal to n, then the system has a unique solution. If rank(**A**) = rank(**A b**) but the rank $< n$, there are an infinite number of solutions. If we denote the rank by r, then r of the unknown variables can be expressed as linear combinations of $n - r$ of the other variables.

Since the rank can be computed easily in MATLAB and we can readily concatenate arrays together, we can use these facts to analyze linear systems with relative ease. If the rank condition is met and the rank is equal to the number of unknowns, the solution can be computed by using left division. Let's illustrate the technique. Consider the system:

$$\begin{aligned} x - 2y + z &= 12 \\ 3x + 4y + 5z &= 20 \\ -2x + y + 7z &= 11 \end{aligned}$$

The coefficient matrix is:

$$A = \begin{pmatrix} 1 & -2 & 1 \\ 3 & 4 & 5 \\ -2 & 1 & 7 \end{pmatrix}$$

We also have:

$$b = \begin{pmatrix} 12 \\ 20 \\ 11 \end{pmatrix}$$

And the augmented matrix is:

$$(\mathbf{A}\ \mathbf{b}) = \begin{pmatrix} 1 & -2 & 1 & 12 \\ 3 & 4 & 5 & 20 \\ -2 & 1 & 7 & 11 \end{pmatrix}$$

The first step is to enter these matrices in MATLAB:

```
>> A = [1 -2 1; 3 4 5; -2 1 7] b = [12; 20; 11]
```

We can create the augmented matrix by using concatenation:

```
>> C = [A b]
C =
     1    -2     1    12
     3     4     5    20
    -2     1     7    11
```

Now let's check the rank of A:

```
>> rank(A)
ans =
     3
```

The rank of the augmented matrix is:

```
>> rank(C)
ans =
     3
```

Since the ranks are the same, a solution exists. We also note that the rank r satisfies $r = n$ since there are three unknown variables. This means the solution is unique. We find it by left division:

```
>> x = A\b
x =
    4.3958
   -2.2292
    3.1458
```

Finding the Inverse of a Matrix and the Pseudoinverse

The inverse of a matrix A is denoted by A^{-1} such that the following relationship is satisfied:

$$AA^{-1} = A^{-1}A = 1$$

Consider the following matrix equation:

$$Ax = b$$

If the inverse of A exists, then the solution can be readily written as:

$$x = A^{-1}b$$

In practice, this is much harder than it looks because computing the inverse of a matrix can be a tedious pain. Luckily MATLAB makes things easy for us by doing all of the tedious work we want to avoid. The inverse of a matrix A can be calculated in MATLAB by writing:

$$\text{inv}(A)$$

The inverse of a matrix does not always exist. In fact, we can use the determinant to determine whether or not the inverse exists. If $\det(A) = 0$, then the inverse does not exist and we say the matrix is *singular.*

Let's get started by calculating a few inverses just to see how easy this is to do in MATLAB. Starting with a simple 2×2 matrix:

$$A = \begin{pmatrix} 2 & 3 \\ 4 & 5 \end{pmatrix}$$

First, let's check the determinant:

```
>> A = [2 3; 4 5]
A =
     2     3
     4     5
>> det(A)
ans =
    -2
```

Since det(A) $\neq 0$, we can find the inverse. MATLAB tells us that it is:

```
>> inv(A)
ans =
   -2.5000    1.5000
    2.0000   -1.0000
```

We can verify that this is the inverse by hand:

$$AA^{-1} = \begin{pmatrix} 2 & 3 \\ 4 & 5 \end{pmatrix}\begin{pmatrix} -5/2 & 3/2 \\ 2 & -1 \end{pmatrix} = \begin{pmatrix} -10/2+6 & 6/2-3 \\ -20/2+10 & 12/2-5 \end{pmatrix} = \begin{pmatrix} 1 & 0 \\ 0 & 1 \end{pmatrix}$$

We aren't going to bother to tell you how to calculate a matrix inverse by hand, that's something you can find in most linear algebra books. Suffice it to say that it's something you want to avoid doing, especially for larger matrices. Let's consider a 4×4 case in MATLAB.

First we create the matrix:

```
>> S = [1 0 -1 2; 4 -2 -3 1; 0 2 -1 1; 0 0 9 8];
```

Checking its determinant we find:

```
>> det(S)
ans =
  -108
```

Since det(S) $\neq 0$, the inverse must exist. MATLAB spits it out for us:

```
>> T = inv(S)
T =
   -0.9259    0.4815    0.4815    0.1111
   -0.6296    0.1574    0.6574    0.0556
   -0.5926    0.1481    0.1481    0.1111
    0.6667   -0.1667   -0.1667         0
```

Finding this by hand would not have been pleasant, and chances are we would have generated some errors. Now let's check that this is in fact the inverse:

```
>> S*T
ans =
    1.0000         0         0              0
    0.0000    1.0000   -0.0000         0
    0.0000   -0.0000    1.0000         0
         0         0         0              1.0000
```

The –0.0000 terms come from some rounding errors and all that. Within computer precision, it looks like we do in fact have the correct inverse. Let's check multiplication the other way:

```
>> T*S
ans =
    1.0000         0         0         0
         0    1.0000         0    0.0000
         0         0    1.0000         0
         0         0    0.0000    1.0000
```

This time, for mysterious reasons left behind the MATLAB curtain that aren't important for those of us just using it for some basics, the result comes out a bit nicer.

Now let's look at how we can solve a system of equations using the inverse. Consider:

$$\begin{aligned} 3x - 2y &= 5 \\ 6x - 2y &= 2 \end{aligned}$$

The coefficient matrix is:

```
>> A = [3 -2; 6 -2]
A =
     3    -2
     6    -2
```

The vector **b** for the system **Ax** = **b** is:

```
>> b = [5;2]
b =
     5
     2
```

First let's check the determinant of A to ensure that the inverse exists:

```
>> det(A)
ans =
     6
```

Since the inverse exists, we can generate the solution readily in MATLAB:

```
>> x = inv(A)*b
x =
   -1.0000
   -4.0000
```

We can only use the method described earlier, multiplying by the inverse of the coefficient matrix to obtain a solution, if the coefficient matrix is square. What this means for the system of equations is that the number of equations equals the number of unknowns. If there are fewer equations than unknowns, the system is called *underdetermined.* This means that the system has an infinite number of solutions. This is because only some of the unknown variables can be determined. The variables that remain unknown can assume any value, hence there are an infinite number of solutions. We take a simple example:

$$x + 2y - z = 3$$
$$5y + z = 0$$

A little manipulation tells us that:

$$z = -5y$$
$$x = 3 - 7y$$

In this system, while we can find values for two of the variables (x and z), the third variable y is undetermined. We can choose any value of y we like, and the system will have a solution.

Another case where an infinite number of solutions exist for a system of equations and unknowns is when det(**A**) = 0.

So what is a poor mathematician to do in such a scenario? Luckily the *pseudoinverse* comes to the rescue. This solution gives the minimum norm solution for real values of the variables. That is, the solution vector **x** is chosen to have the smallest norm such that the components of **x** are real. Let's consider a linear system of equations:

$$3x + 2y - z = 7$$
$$4y + z = 2$$

Obviously this system has an infinite number of solutions. We enter the data:

```
>> A= [3 2 -1; 0 4 1]; b = [7;2];
>> C = [A b]
C =
     3     2    -1     7
     0     4     1     2
```

Now we compute the ranks:

```
>> rank(A)
ans =
     2
>> rank(C)
ans =
     2
```

Since these ranks are equal, a solution exists. We can have MATLAB generate a solution using left division:

```
>> x = A\b
x =
    2.0000
    0.5000
         0
```

MATLAB has generated a solution by setting one of the variables (z in this case) to zero. This is typically what it does in cases like these, if you try to generate a solution using left division. The solution is valid of course, but remember it only holds when $z = 0$, and z can be anything.

We can also solve the system using the *pseudoinverse*. We do this by typing:

```
>> x = pinv(A)*b
x =
    1.6667
    0.6667
   -0.6667
```

MATLAB uses the Moore-Penrose pseudoinverse to calculate pinv.

Reduced Echelon Matrices

The MATLAB function rref(A) uses Gauss-Jordan elimination to generate the reduced row echelon form of a matrix A. A simple example you can verify with a hand calculation:

```
>> A = [1 2; 4 7]
A =
     1     2
     4     7
>> rref(A)
ans =
     1     0
     0     1
```

Now let's consider an example you're unlikely to calculate by hand. A *magic* matrix is an $n \times n$ matrix. The components of the matrix range from 1 to n^2 and the sum of the elements in a column is equal to the sum of the elements in a row. Trying to generate one of these by hand might cause a brain hemorrhage, but MATLAB does it for us with the magic(*n*) command. For example:

```
>> A = magic(5)
A =
    17    24     1     8    15
    23     5     7    14    16
     4     6    13    20    22
    10    12    19    21     3
    11    18    25     2     9
```

Let's check the sum of the columns to see that they're equal:

```
>> sum(A)
ans =
    65    65    65    65    65
```

Using Gauss-Jordan elimination on this matrix to find the reduced row echelon form would be a bit too much work for me to tackle. So I am going to have MATLAB tell us what it is:

```
>> rref(A)
ans =
     1     0     0     0     0
     0     1     0     0     0
     0     0     1     0     0
     0     0     0     1     0
     0     0     0     0     1
```

Well how about that. We got the identity again. Let's try a larger matrix:

```
>> magic(8)
ans =
    64     2     3    61    60     6     7    57
     9    55    54    12    13    51    50    16
    17    47    46    20    21    43    42    24
    40    26    27    37    36    30    31    33
    32    34    35    29    28    38    39    25
    41    23    22    44    45    19    18    48
    49    15    14    52    53    11    10    56
     8    58    59     5     4    62    63     1
```

This matrix is too big for most of us to tackle, but MATLAB finds:

```
>> rref(magic(8))
ans =
     1     0     0     1     1     0     0     1
     0     1     0     3     4    -3    -4     7
     0     0     1    -3    -4     4     5    -7
     0     0     0     0     0     0     0     0
     0     0     0     0     0     0     0     0
     0     0     0     0     0     0     0     0
     0     0     0     0     0     0     0     0
     0     0     0     0     0     0     0     0
```

Matrix Decompositions

MATLAB can be used to quickly generate the LU, QR, or SVD decompositions of a matrix. In this section, we will take a look at LU decomposition and see how to use it to solve a linear system of equations in MATLAB. We can find the LU decomposition of a matrix A by writing:

[L, U] = lu(A)

For example, let's find the LU decomposition of:

$$A = \begin{pmatrix} -1 & 2 & 0 \\ 4 & 1 & 8 \\ 2 & 7 & 1 \end{pmatrix}$$

We enter the matrix and find:

```
>> A = [-1 2 0; 4 1 8; 2 7 1];
>> [L, U] = lu(A)
L =
   -0.2500    0.3462    1.0000
    1.0000         0         0
    0.5000    1.0000         0
U =
    4.0000    1.0000    8.0000
         0    6.5000   -3.0000
         0         0    3.0385
```

We can use the LU decomposition to solve a linear system. Suppose that A was a coefficient matrix for a system with

$$b = \begin{pmatrix} 12 \\ -8 \\ 6 \end{pmatrix}$$

The solution can be generated with two left divisions:

x = U\(L\b)

We find:

```
>> x = U\(L\b)
x =
   -6.9367
    2.5316
    2.1519
```

Consider the system:

$$\begin{aligned} 3x + 2y - 9z &= -65 \\ -9x + 5y + 2z &= 16 \\ 6x + 7y + 3z &= 5 \end{aligned}$$

We enter the system in MATLAB:

```
>> A = [3 2 -9; -9 -5 2; 6 7 3]; b = [-65; 16; 5];
```

Now let's find the LU decomposition of A:

```
>> [L, U] = lu(A)
L =
   -0.3333    0.0909    1.0000
    1.0000         0         0
   -0.6667    1.0000         0
U =
   -9.0000   -5.0000    2.0000
         0    3.6667    4.3333
         0         0   -8.7273
```

Now we use these matrices together with left division to generate the solution:

```
>> x = U\(L\b)
x =
    2.0000
   -4.0000
    7.0000
```

Quiz

1. Find the magnitude of the vector $\boldsymbol{A} = (-1 \quad 7 \quad 3 \quad 2)$.
2. Find the magnitude of the vector $\boldsymbol{A} = (-1 + i \quad 7i \quad 3 \quad -2-2i)$.
3. Consider the numbers 1, 2, 3. Enter these as components of a column vector and as components of a row vector.
4. Given **A** = [1; 2; 3]; B = [4; 5; 6];, find the array product of the two vectors.
5. What command would create a 5×5 matrix with ones on the diagonal and zeros everywhere else?
6. Consider the two matrices $A = \begin{pmatrix} 8 & 7 & 11 \\ 6 & 5 & -1 \\ 0 & 2 & -8 \end{pmatrix}, B = \begin{pmatrix} 2 & 1 & 2 \\ -1 & 6 & 4 \\ 2 & 2 & 2 \end{pmatrix}$ and compute their array product and matrix product.
7. Suppose that $A = \begin{pmatrix} 1 & 2 & 3 \\ 4 & 5 & 6 \\ 7 & 8 & 9 \end{pmatrix}$. Use it to create $B = \begin{pmatrix} 7 & 8 & 9 \\ 7 & 8 & 9 \\ 4 & 5 & 6 \end{pmatrix}$

8. Find a solution to the following set of equations:

$$x + 2y + 3z = 12$$
$$-4x + y + 2z = 13$$
$$9y - 8z = -1$$

What is the determinant of the coefficient matrix?

9. Does a solution to the following system exist? What is it?

$$x - 2y + 3z = 1$$
$$x + 4y + 3z = 2$$
$$2x + 8y + z = 3$$

10. Use LU decomposition to find a solution to the system:

$$x + 7y - 9z = 12$$
$$2x - y + 4z = 16$$
$$x + y - 7z = 16$$

CHAPTER 3

Plotting and Graphics

Plotting is one of the most useful applications of a math package used on the computer, and MATLAB is no exception to this rule. Often we need to visualize functions that are too hard to graph "by hand" or to plot experimental or generated data. In this chapter we will introduce the commands and techniques used in MATLAB to accomplish these kinds of tasks.

Basic 2D Plotting

Let's start with the most basic type of plot we can create, the graph of a function of one variable. Plotting a function in MATLAB involves the following three steps:

1. Define the function
2. Specify the range of values over which to plot the function
3. Call the MATLAB plot(x, y) function

When specifying the range over which to plot the function, we must also tell MATLAB what increment we want it to use to evaluate the function. It doesn't take a rocket scientist to figure out that using smaller increments will result in plots with a smoother appearance. If the increment is smaller, MATLAB will evaluate the function at more points. But it's generally not necessary to go that small. Let's look at a simple example to see how this works.

Let's plot the function $y = \cos(x)$ over the range $0 \le x \le 10$. To start, we want to define this interval and tell MATLAB what increment to use. The interval is defined using square brackets [] that are filled in the following manner:

```
[ start : interval : end ]
```

For example, if we want to tell MATLAB to plot over $0 \le x \le 10$ with an interval of 0.1, we type:

```
[0:0.1:10]
```

To assign this range to a variable name, we use the assignment operator. We also do this to tell MATLAB what the dependent variable is and what function we want to plot. Hence to plot $y = \cos(x)$, we enter the following commands:

```
>> x = [0:0.1:10];
>> y = cos(x)
```

Notice that we ended each line with semicolons. Remember, this suppresses MATLAB output. It's unlikely you would want MATLAB to spit out all the x values over the interval onto the screen, so we use the semicolon to prevent this. Now we can plot the function. This is done by entering the following command:

```
>> plot(x, y)
```

After typing the plot command, hit the ENTER key. After a moment MATLAB will open a new window on the screen with the caption *Figure 1.* The plot is found in this window. For the example we used, we obtain the plot shown in Figure 3-1.

Now what about the increment? Suppose that we make the increment ten times as large, so we set it to 1. This is done by entering the following command:

```
>> x = [0:1:10];
```

If we try to plot again, we get an error message:

```
>> plot(x, y);
??? Error using ==> plot
Vector must be the same lengths.
```

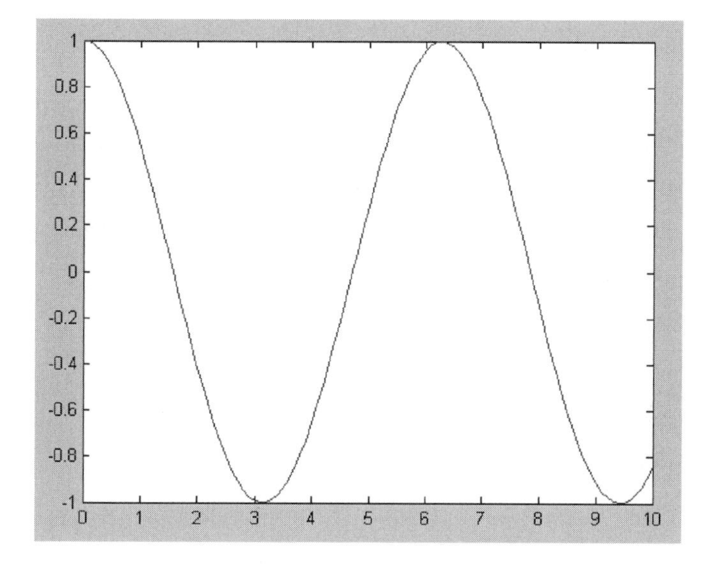

Figure 3-1 A plot of $y = \cos(x)$ generated by MATLAB for $0 \leq x \leq 10$

We had already made the definition $y = \cos(x)$, yet MATLAB can't plot it. What happened here? Turns out we need to tell MATLAB to reevaluate the function for the number of points and values that we gave it when we redefined x. In other words, the correct course of action is to reenter all of the commands all over again:

```
>> y = cos(x)
y =
    Columns 1 through 10
      1.000    0.5403    -0.4161    -0.9900    -0.6536    0.2837    0.9602    0.7539    -0.1455    -0.9111
    Column 11
     -08391
>> plot(x, y)
```

A quick aside—notice that we left off the semicolon after our definition of y—so MATLAB outputs the evaluation of cos(x) at each point. If you have a large number of points, you can see that this probably would be undesirable.

OK back to the plot. When we plot with a larger increment, the plot is less accurate. Look at what MATLAB generates for $y = \cos(x)$ in Figure 3-2, where we used an increment size of 1.

The plot of the function in this case is choppy. Let's try going the other way. We will make the increment ten times smaller than our original attempt, setting it to 0.01.

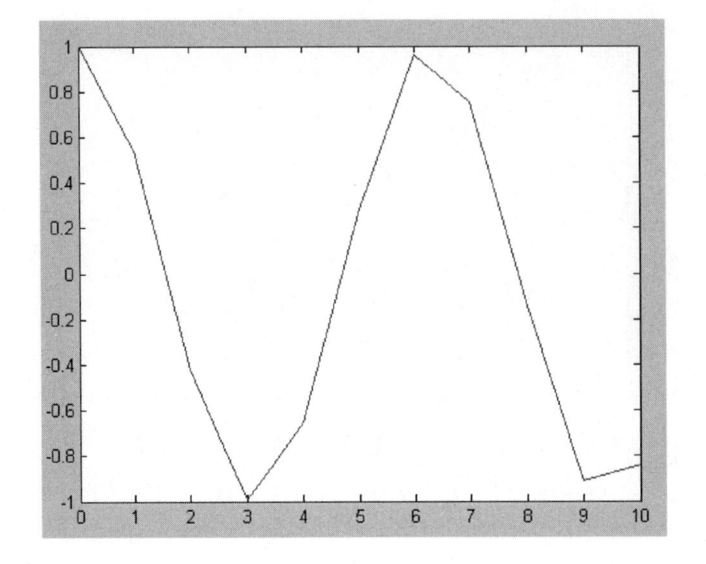

Figure 3-2 A plot of the cosine function with a larger increment

Remember we need to redefine *y* again, therefore the commands we need to type in are:

```
>> x = [0:0.01:10];
>> y = cos(x);
>> plot(x, y)
```

This time we get a very nice smooth rendition of $y = \cos(x)$, this is shown in Figure 3-3.

OK now we know how to get a straightforward plot on the screen. The next thing you might want to do is generate a plot that had the axes labeled. This can be done using the *xlabel* and *ylabel* functions. These functions can be used with a single argument, the label you want to use for each axis enclosed in quotes. Place the *xlabel* and *ylabel* functions separated by commas on the same line as your plot command. For example, the following text generates the plot shown in Figure 3-4:

```
x=[0:0.01:10];
y=cos(x);
plot(x,y), xlabel('x'), ylabel('cos(x)');
```

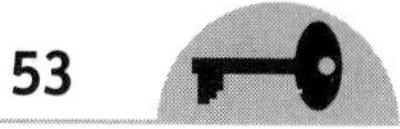

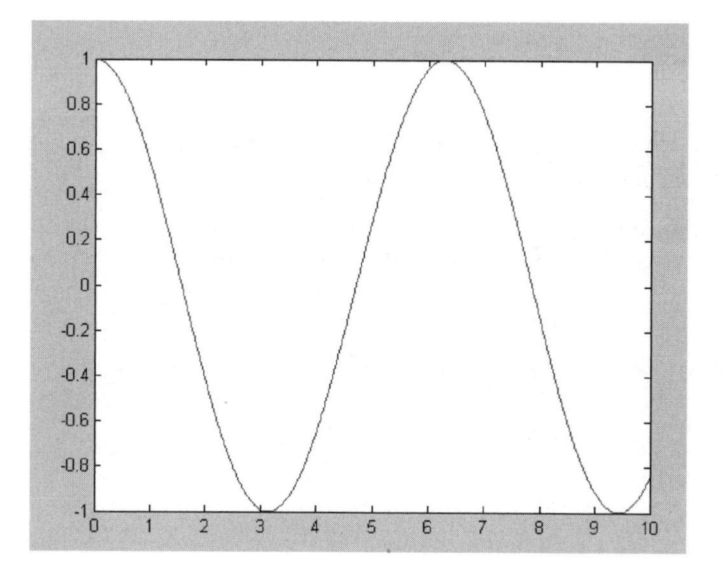

Figure 3-3 A plot of $y = \cos(x)$ with a smaller increment

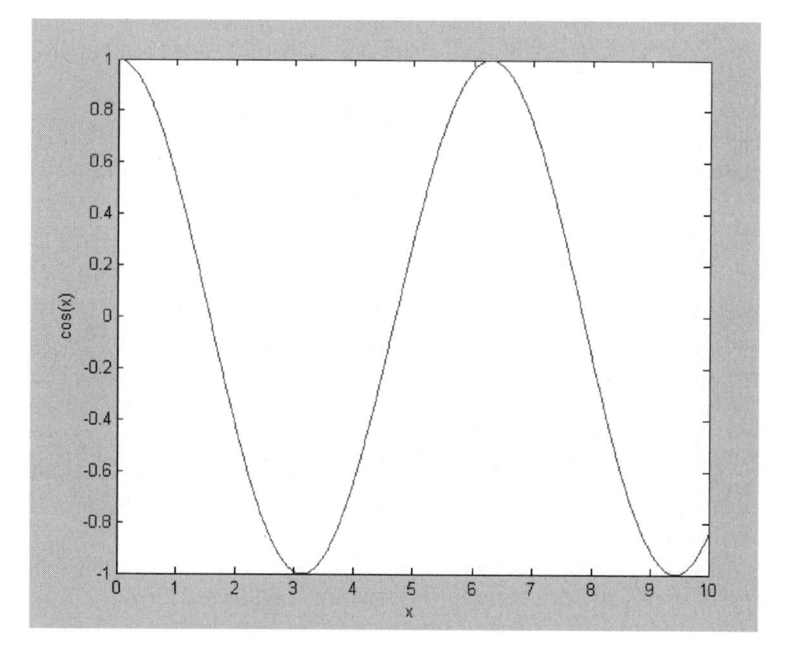

Figure 3-4 Sprucing up the plot with axis labels

More 2D Plotting Options

OK so at this point we basically know how to spit out a generic plot of a given function. Let's look at some other options we might consider with plots. If you're going to use a plot in a presentation or homework assignment, you might want to give the plot a title. MATLAB allows you to do this by using the *title* command which takes a character string enclosed in single quotes. The title is printed just above the plot. Let's say that we were plotting some force data that happened to vary as $f(t)$ $e^{-2t}\sin t$ for $0 \leq t \leq 4$, where t is time in seconds with data at intervals of 0.02 seconds. We want to title the plot "Damped Spring Forcing." How do we do it?

The first step is to define our interval. We do this in the usual way, calling our independent variable t instead of x.

```
>> t = [0:0.02:1];
```

Now let's define our function. This should be pretty straightforward:

```
>> f = exp(-t)*sin(t);
```

However, if you try this, you'll notice we get an error message. MATLAB tells us that

```
??? Error using ==> mtimes
Inner matrix dimensions must agree.
```

So how do we get around this? One way is to use the *fplot* function instead. The function fplot gets around our choice of interval used to generate the plot, and instead decides the number of plotting points to use for us. Generally, fplot will allow you to generate the most accurate plots possible, but it also helps us get around errors like these. The formal call to fplot goes like fplot ('function string,' [*x*start, *x*end]). The argument *function string* tells fplot the function you want to plot while *xstart* and *xend* define the range over which to display the plot. This is pretty straightforward, let's see how it works out in the current example.

We can do it all in one go by typing the following command and hitting the ENTER key:

```
>> fplot('exp(-2*t)*sin(t)',[0, 4];
```

MATLAB quickly produces the plot shown in Figure 3-5.

If we want to add labels and a title to the plot, we can follow the same procedure used with plot(x,y). Let's try it again this time adding our title "Damped Spring Forcing" and labeling our axes:

```
fplot('exp(-2*t)*sin(t)',[0,4]), xlabel('t'), ylabel('f(t)'),
title('Damped Spring Forcing')
```

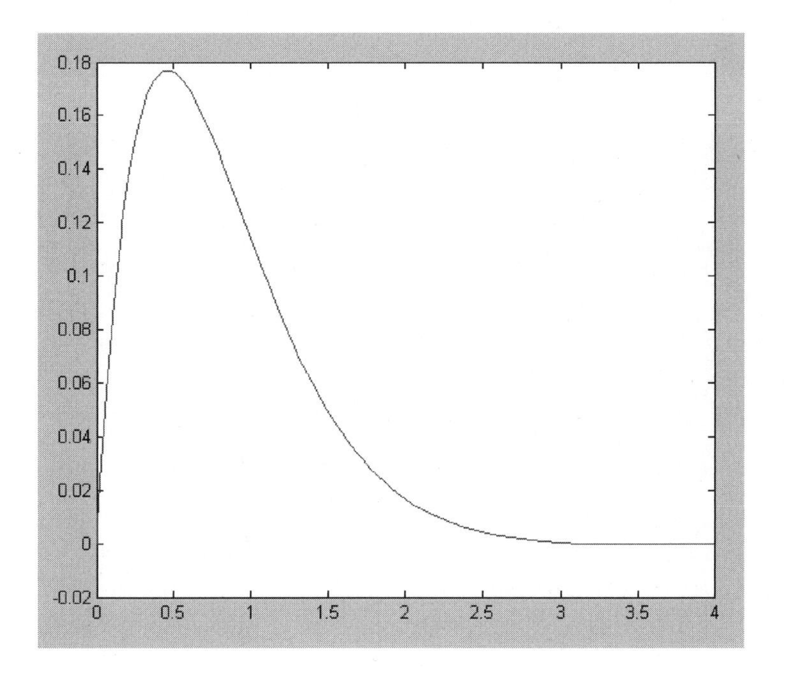

Figure 3-5 The function $f(t)$ $e^{-2t}\sin t$ generated using the *fplot* command

This produces the labeled plot shown in Figure 3-6.

Well this provided a nice intro to the fplot command, but it turns out we had a typo in our command multiplying the exponential and trig function which generated the error. We had typed:

```
>> f = exp(-t)*sin(t);

??? Error using ==> mtimes
Inner matrix dimensions must agree.
```

The correct way to type this in MATLAB is to include a period prior to the multiplication symbol. Confused? Let's show the correct way to type this function.

```
>> t = [0:0.01:4];
>> f = exp(-2*t).*sin(t);
>> plot(t, f)
```

This time, there is no error and the function plots correctly. Therefore, when generating a function, which is formed by the product of two or more other functions, be sure to tell MATLAB we're multiplying two matrices by including a '.' character.

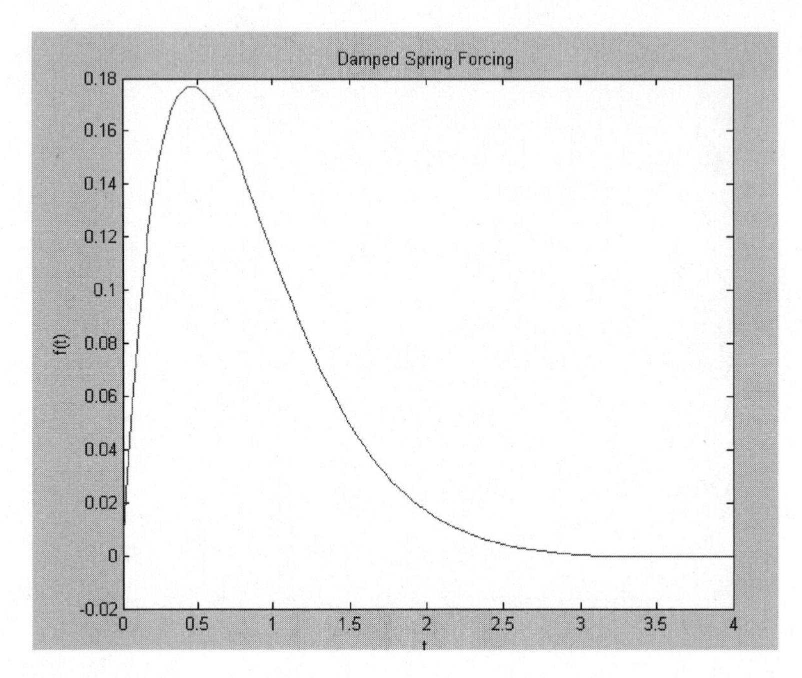

Figure 3-6 A labeled plot generated with the *fplot* command

For the current example we replaced `>> f = exp(-t)*sin(t);` with `y = exp(-1.2*x).*sin(10*x + 5);`. The difference is that in the second case, we used matrix multiplication (indicated by typing .*).

Well we are back to the old plot(x, y) command. What are some other ways we can spruce up your basic two-dimensional plot? One way is to add a grid to the plot. This is done by adding the phrase *grid on* to your plot statement. For the next example, we will plot $y = \tanh(x)$ over the range $-6 \le x \le 6$ with a grid display. First we define our interval:

```
>> x = [-6:0.01:6];
```

Next, we define the function:

```
>> y = tanh(x);
```

The plot command looks like this, and produces the plot shown in Figure 3-7.

```
Plot(x,y),grid on
```

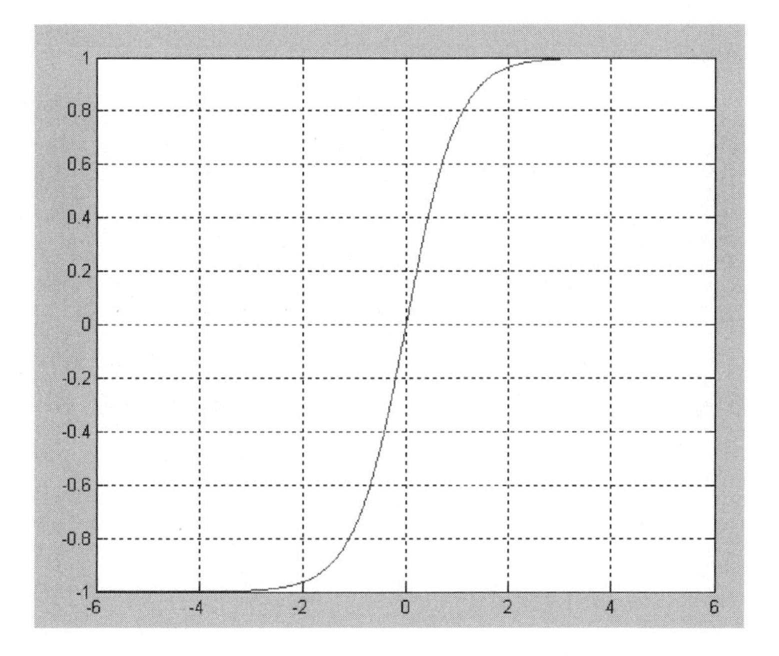

Figure 3-7 A plot made with the *grid on* command

The Axis Commands

MATLAB allows you to adjust the axes used in two-dimensional plots in the following way. If we add *axis square* to the line containing the plot command, this will cause MATLAB to generate a square plot. If we type *axis equal*, then MATLAB will generate a plot that has the same scale factors and tick spacing on both axes.

Let's return to the example $y = \tanh(x)$, which we plotted in Figure 3-7. If you run this plot with axis square, you will get the same plot that we did using the default settings. But suppose that we typed:

```
>> plot(x,y),grid on, axis equal
```

In this case, we get the plot shown in Figure 3-8. Notice that the spacing used for the y axis in Figure 3-7 and Figure 3-8 are quite different. In the first case, the spacing used on the vertical or y axis is different than the spacing used on the x axis. In contrast, in Figure 3-8, the spacing is identical.

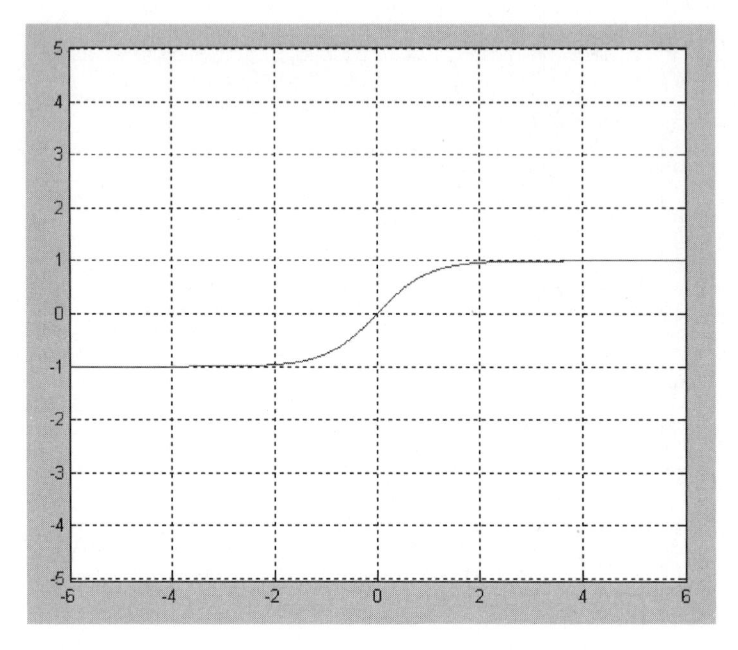

Figure 3-8 Plotting $y = \tanh(x)$ using the axis equal option

As you can see from this dramatic example, we can use the axis command to generate plots that differ quite a bit in appearance. Hence we can use the command to play with different plot styles and select what we need for the particular application. To let MATLAB set the axis limits automatically, type *axis auto*. This isn't necessary, of course, unless you've been playing with the options described here.

Showing Multiple Functions on One Plot

In many cases, it is necessary to plot more than one curve on a single graph. The procedure used to do this in MATLAB is fairly straightforward. Let's start by showing two functions on the same graph. In this case let's plot the following two functions over $0 \le t \le 5$:

$$f(t) = e^{-t}$$
$$g(t) = e^{-2t}$$

We will differentiate between the two curves by plotting g with a dashed line. Following the usual procedure, we first define our interval:

```
>> t = [0:0.01:5];
```

Next, we define the two functions:

```
>> f = exp(-t);
>> g = exp(-2*t);
```

To plot multiple functions, we simply call the plot(x, y) command with multiple pairs x, y defining the independent and dependent variables used in the plot in pairs. This is followed by a character string enclosed in single quotes to tell us what kind of line to use to generate the second curve. In this case we have:

```
>> plot(t,f,t,g,'--')
```

This tells MATLAB to generate plots of $f(t)$ and $g(t)$ with the latter function displayed as a dashed line. Note that while we can't show it in the book, MATLAB displays each curve with a unique color. The result is shown in Figure 3-9.

MATLAB has four basic line types that can be defined in a plot. These are, along with the character strings, used to define them in the plot command:

- Solid line '-'
- Dashed line '--'

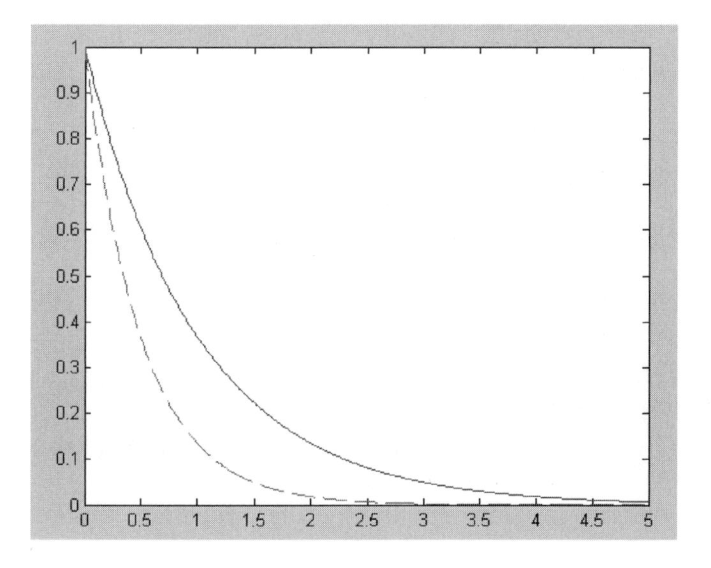

Figure 3-9 Plotting two curves on the same graph

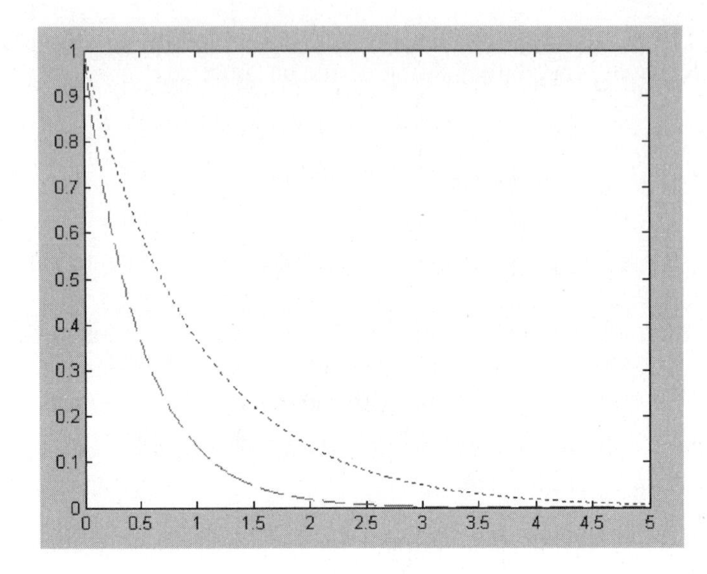

Figure 3-10 Using a dotted line to represent $f(t) = e^{-t}$ and a dashed line to represent $g(t) = e^{-2t}$

- Dash-dot line '-.'
- Dotted line ':'

Let's generate the same graph as in Figure 3-9 making the curve $f(t) = e^{-t}$ appear with a dotted line. The command is

```
plot(t,f,':',t,g,'--')
```

This generates the plot shown in Figure 3-10.

If you want to plot all curves using solid lines and simply differentiate them by their colors, just leave off the character string specifying the curve type. The plot will be generated using solid lines, which is the default.

Adding Legends

A professionally done plot often has a legend that lets the reader know which curve is which. In the next example, let's suppose that we are going to plot two potential energy functions that are defined in terms of the hyperbolic trig functions sinh(x) and cosh(x) for $0 \le x \le 2$. First we define x:

```
>> x = [0:0.01:2];
```

Now we define our two functions. There is nothing magical about calling a function *y* or anything else in MATLAB, so let's call the second function *z*. So we have

```
>> y = sinh(x);
>> z = cosh(x);
```

The *legend* command is simple to use. Just add it to the line used for the plot(*x*, *y*) command and add a text string enclosed in single quotes for each curve you want to label. In our case we have:

```
legend('sinh(x)','cosh(x)')
```

We just add this to the plot command. For this example, we include *x* and *y* labels as well, and plot the curves using a solid line for the first curve and a dot-dash for the second curve:

```
>> plot(x,y,x,z,'-.'),xlabel('x'),ylabel('Potential'),legend('sinh(x)','cosh(x)')
```

The plot that results is shown in Figure 3-11. The legend didn't originally show up where it is in the figure, and it probably won't do so on your system either. To move the legend to a more favorable position that might be better for printing or display, just hold the mouse pointer over the legend and drag it to the location where you want it to display.

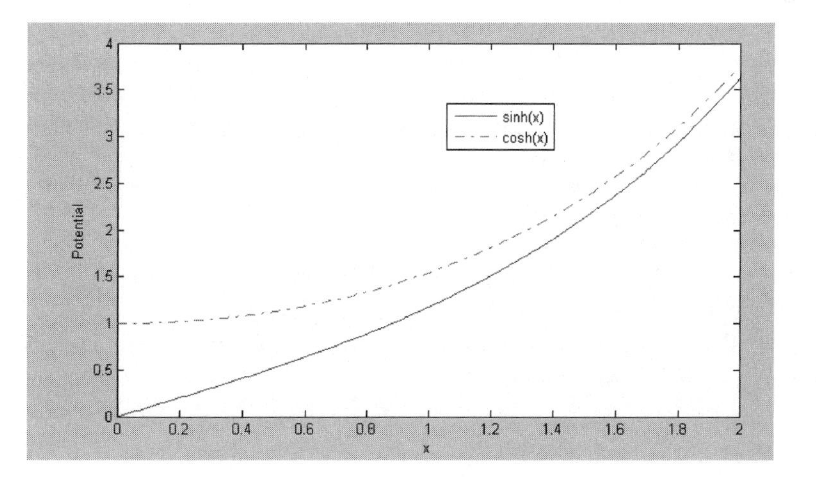

Figure 3-11 A plot of two curves that includes a legend

Setting Colors

The color of each curve can be set automatically by MATLAB or we can manually select which color we want. This is done by enclosing the appropriate letter assigned to each color used by MATLAB in single quotes immediately after the function to be plotted is specified. Let's illustrate with an example.

Let's plot the hyperbolic sine and cosine functions again. This time we'll use a different interval for our plot, we will take $-5 \le x \le 5$. So we define our data array as

```
>> x = [-5:0.01:5];
```

Now we redefine the functions. Remember if we don't do this and we're in the same session of MATLAB, the program is going to think that the functions are defined in terms of the previous x we had used. So now we type:

```
>> y = sinh(x);
>> z = cosh(x);
```

Now we will generate the plot representing y with a red curve and z with a blue curve. We do this by following our entries for y and z in the *plot* function by the character strings 'r' and 'b' respectively. The command looks like this:

```
>> plot(x,y,'r',x,z,'b')
```

Try this on your own system to see the plot it generates. Now, it is possible to set more than one option for each curve. So let's use the colors red and blue for the curves, and set the cosh function (the blue curve) to draw with a dashed line. This is done in the following way, enclosing all of the plot options for the selected curve within the same set of quotes:

```
>> plot(x,y,'r',x,z,'b--')
```

This gives us the plot shown in Figure 3-12. You can't see it in the black and white image, but the dashed line prints on screen as a blue curve.

MATLAB gives the user eight basic color options for drawing curves. These are shown with their codes in Table 3-1.

Table 3-1 MATLAB specifiers for selecting plot colors.

Color	Specifier
White	w
Black	k
Blue	b
Red	r
Cyan	c
Green	g
Magenta	m
Yellow	y

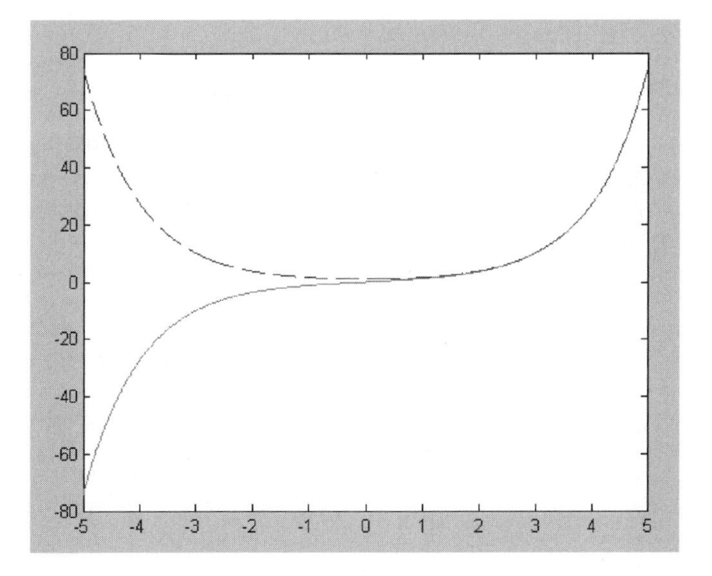

Figure 3-12 A plot generated setting colors and line types with the same command

Setting Axis Scales

Let's take another look at the *axis* command and see how to set the plot range. This is done by calling axis in the following way:

```
axis ( [xmin xmax ymin ymax] )
```

Suppose that we want to generate a plot of $y = \sin(2x + 3)$ for $0 \le x \le 5$. We might consider that the function ranges over $-1 \le y \le 1$. We can set the y axis to only show these values by using the following sequence of commands:

```
>> x = [0:0.01:5];
>> y = sin(2*x + 3);
>> plot(x,y), axis([0 5 -1 1])
```

This generates the plot shown in Figure 3-13.

Now let's make a plot of $y = e^{-3/2x} \sin(5x + 3)$. First we try $0 \le x \le 5$, $-1 \le y \le 1$.

```
>> y = exp(-1.5*x).*sin(5*x+3);
>> plot(x,y), axis([0 5 -1 1])
```

This generates the plot shown in Figure 3-14. As you can see from the figure, the range used for the y axis could be adjusted.

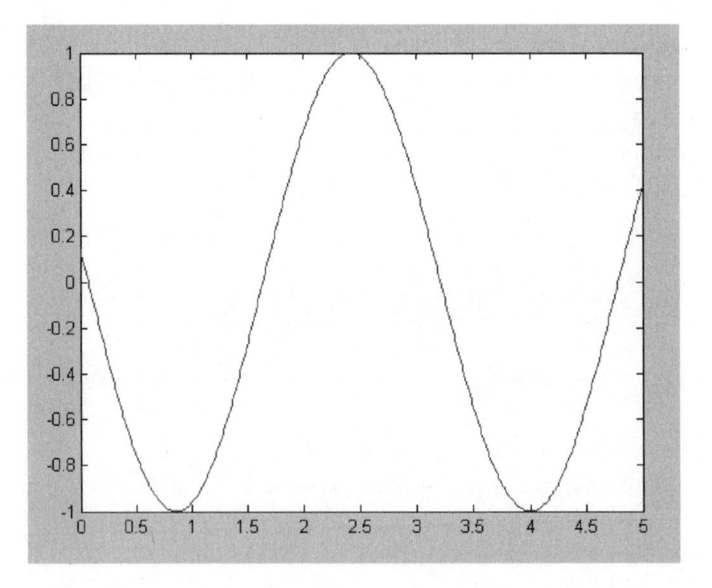

Figure 3-13 A plot generated manually setting the limits on the x and y axes for a plot of $y = \sin(2x + 3)$ for $0 \le x \le 5$

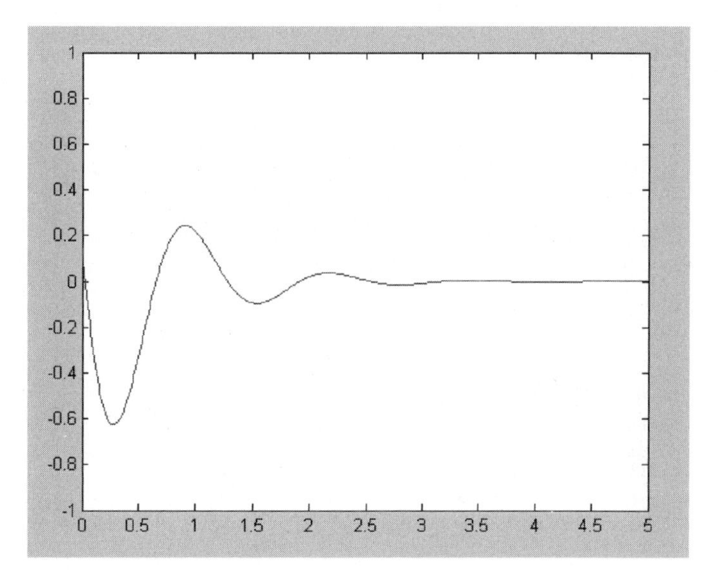

Figure 3-14 A plot of $y = e^{-3/2x}\sin(5x + 3)$. First we try $0 \le x \le 5$, $-1 \le y \le 1$

Let's try adjusting the range of y values on the plot, so that $-0.7 \le y \le 0.3$. We do this by adjusting the axis command as follows:

```
>> plot(x,y), axis([0 5 -0.7 0.3])
```

This gives us a much tighter view of the graph, as shown in Figure 3-15.

We aren't restricted to plot only over the entire set of values of x we use to generate a function. To see what we mean by this, let's generate a couple of plots of $y = \sin^2(5x)$. First as an aside, let's make a note of how one would square the sin function in MATLAB. If you type:

```
>> y = sin(5*x)^2
```

MATLAB is going to lash out at you:

```
??? Error using ==> mpower
```

Matrix must be square.

The correct way to square the sin function is to use the arraywise power notation, which uses $A.\,\hat{}B$ to represent A^B. Hence the following command will work

```
>> y = sin(5*x).^2;
```

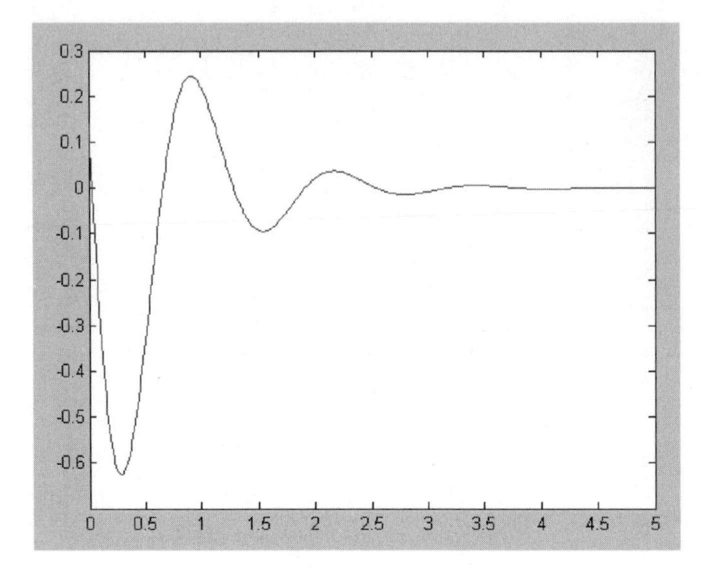

Figure 3-15 The plot of $y = e^{-3/2x} \sin(5x + 3)$ with $-0.7 \le y \le 0.3$, generated by using axis([0 5 –0.7 0.3])

This squares each element of the array, instead of the array as a whole. Now let's plot it using the automatic settings. If we just type plot(*x, y*), then MATLAB generates the plot shown in Figure 3-16.

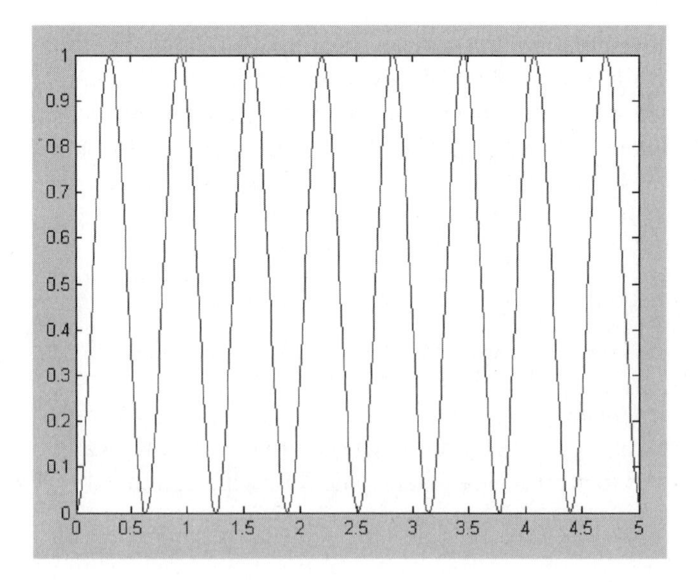

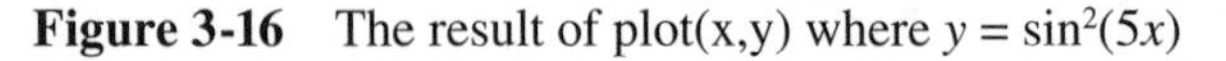
Figure 3-16 The result of plot(x,y) where $y = \sin^2(5x)$

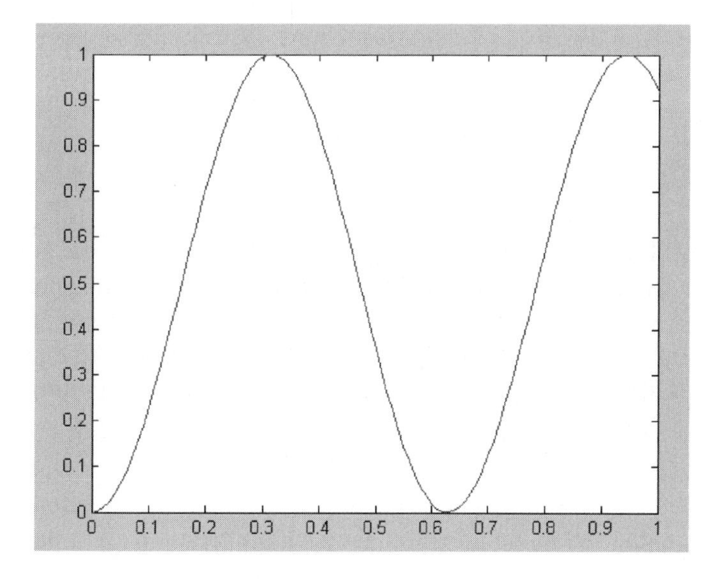

Figure 3-17 A plot of $y = \sin^2(5x)$ again, this time with $0 \le x \le 1$

Suppose that we only want to look at the plot over a restricted set of x values. For example, we can set $0 \le x \le 1$ by typing:

```
>> plot(x,y), axis([0 1 0 1])
```

This generates the plot shown in Figure 3-17.

At this point you should have a handle on the basics needed to generate plots in MATLAB. Now let's consider putting two or more plots in the same figure.

Subplots

A *subplot* is one member of an array of plots that appears in the same figure. The subplot command is called using the syntax subplot(m, n, p). Here m and n tell MATLAB to generate a plot array with m rows and n columns. Then we use p to tell MATLAB where to put the particular plot we have generated. As always, these ideas are best illustrated with an example.

Each plot created with the subplot command can have its own characteristics. For our first example, we will show $y = e^{-1.2x}\sin(20x)$ and $y = e^{-2x}\sin(20x)$ side by side.

In both cases, we will set $0 \le x \le 5$ and $-1 \le y \le 1$. First we define the values used in our domain, define the first function, and then make a call to subplot:

```
>> x = [0:0.01:5];
>> y = exp(-1.2*x).*sin(20*x);
>> subplot(1,2,1)
```

By passing (1, 2, 1) to subplot, we have told MATLAB that we are going to create an array with 2 panes and 1 row, and that this particular plot will appear in the first pane. Panes are numbered in the usual way, moving from left to right, so this plot will appear in the left pane. At this point, MATLAB has generated the first pane in the figure, but hasn't placed anything in it. This is illustrated in Figure 3-18.

Now we call the plot command:

```
>> plot(x,y),xlabel('x'),ylabel('exp(-1.2x)*sin(20x)'),axis([0 5 -1 1])
```

If we look at the figure, the function has been plotted in the first pane, as shown in Figure 3-19.

Figure 3-18 The result of the first call to subplot

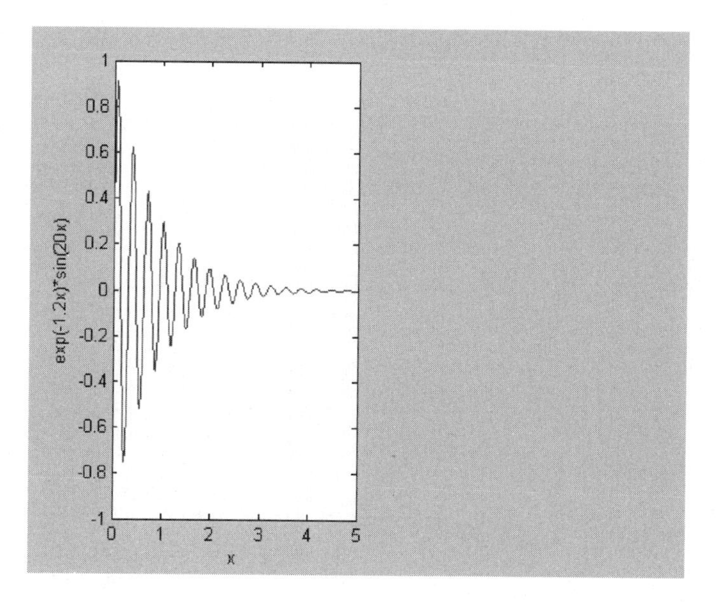

Figure 3-19 A glance at the MATLAB output after our first calls to subplot and plot

With the first plot created, we can move on to generating the second plot. First we define the function

```
>> y = exp(-2*x).*sin(20*x);
```

(Aside—remember to include the period before * when multiplying two functions together). Now we call subplot, this time telling MATLAB to place this function in the *second* pane:

```
>> subplot(1,2,2)
```

MATLAB hasn't put any data in there yet—remember we still have to call the plot command. The figure now looks like that shown in Figure 3-20.

Now we plot it:

```
>> plot(x,y),xlabel('x'),ylabel('exp(-2x)*sin(20x)'),axis([0 5 -1 1])
```

The result, two side-by-side plots, is shown in Figure 3-21.

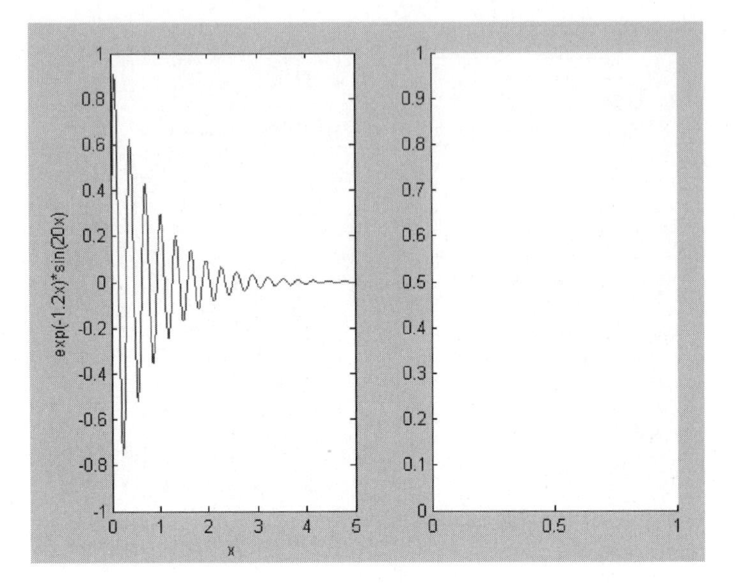

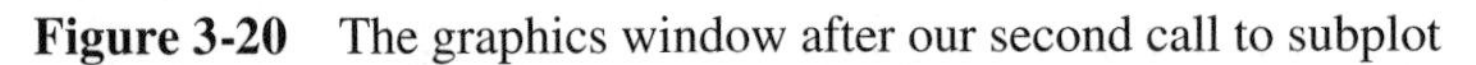
Figure 3-20 The graphics window after our second call to subplot

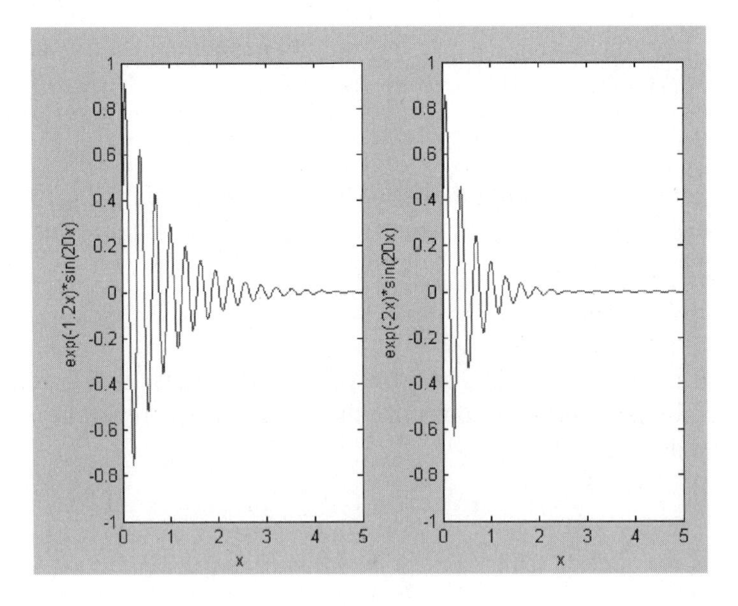

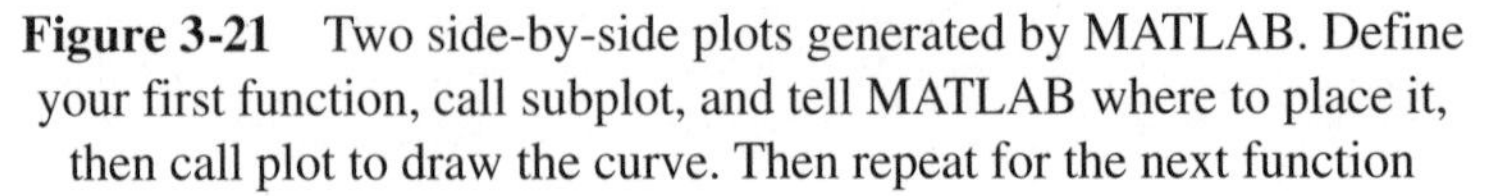
Figure 3-21 Two side-by-side plots generated by MATLAB. Define your first function, call subplot, and tell MATLAB where to place it, then call plot to draw the curve. Then repeat for the next function

Overlaying Plots and linspace

Let's suppose that we plot a function, and then decide that we want the plot of a second function to appear on the same graph. We can do this with two calls to the plot command by telling MATLAB to *hold on.*

In the following example we will generate plots of cos(x) and sin(x) and place them on the same graphic. First, let's learn a new command that can be used to generate a set of x data. This can be done using the *linspace* command. It can be called in one of two ways. If we write:

```
x = linspace(a,b)
```

MATLAB will generate a line (or row vector) of 100 uniformly spaced points from a to b. If instead we write

```
x = linspace(a,b,n)
```

Then MATLAB will create a line of n uniformly spaced points from a to b. Now let's use this tool to plot cos(x) and sin(x). We define a set of 100 linearly spaced points from 0 to 2π by entering the following command:

```
>> x = linspace(0,2*pi);
```

Now let's plot cos(x);

```
>> plot(x,cos(x))
```

We get the graphic shown in Figure 3-22.

If now type:

```
>> plot(x,sin(x))
```

MATLAB just overwrites our previous output. Now the graphics window has the plot shown in Figure 3-23.

A quick detour—notice that even though we defined our range of x values to be $0 \leq x \leq 2\pi$, MATLAB has carried the graph out a bit further than where the function has been calculated. We can fix that up by including the axis command in our call to plot(x,y):

```
>> plot(x,sin(x)),axis([0 2*pi -1 1])
```

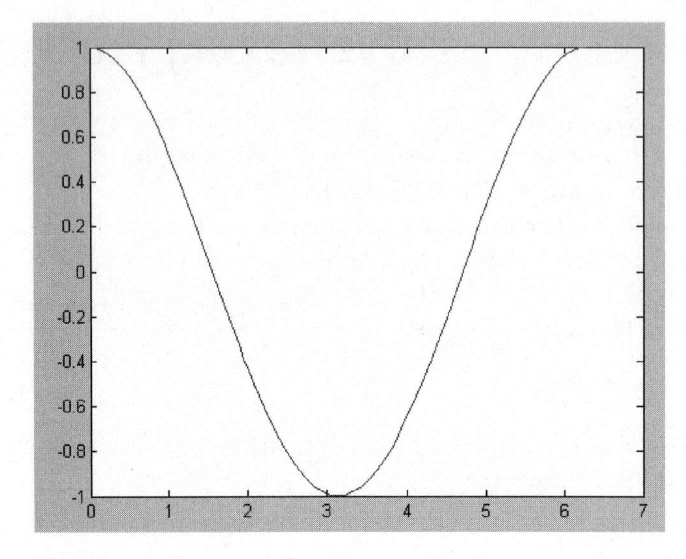

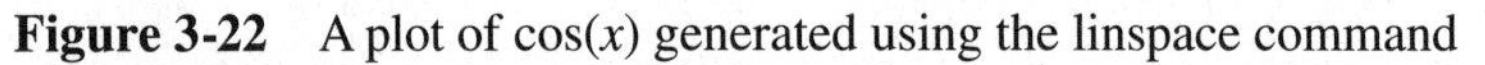
Figure 3-22 A plot of cos(x) generated using the linspace command

This generates the nicer plot shown in Figure 3-24.

Returning to our dilemma, let's say we have plotted cos(x) and want to overlay sin(x) on the same graphic. We can do this with the following sequence of commands:

```
>> x = linspace(0,2*pi);
>> plot(x,cos(x)),axis([0 2*pi -1 1])
```

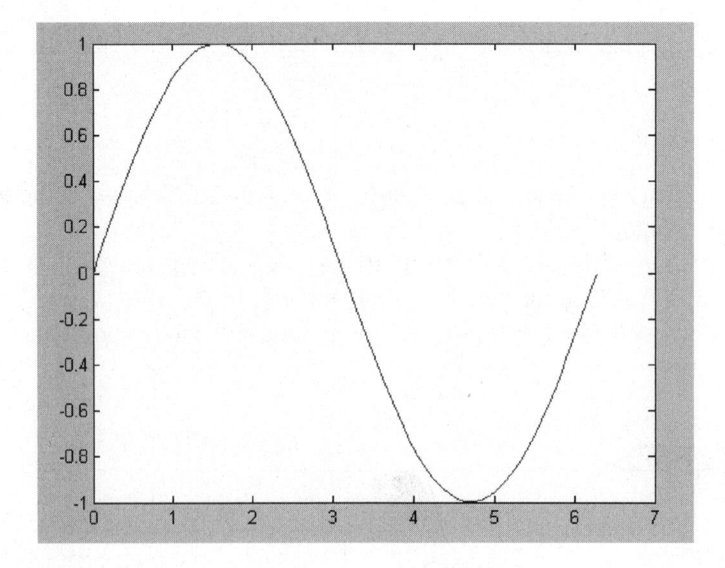

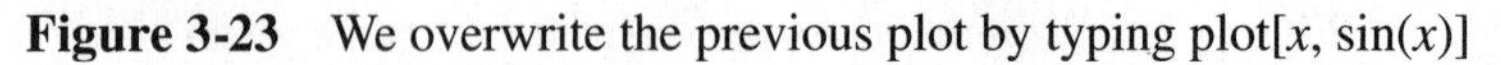
Figure 3-23 We overwrite the previous plot by typing plot[x, sin(x)]

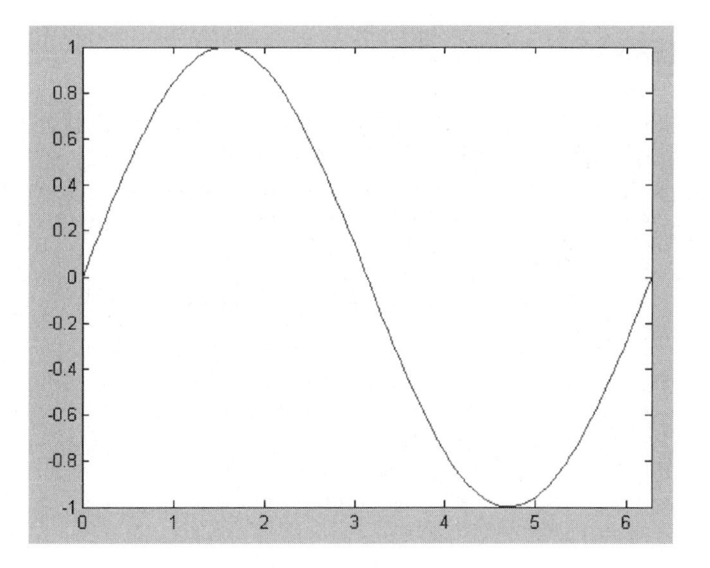

Figure 3-24 Now we fix the plot by calling axis

```
>> hold on
>> plot(x, sin(x)), axis ([0 2*pi -1 1])
```

The result, shown in Figure 3-25, displays both curves on the same plot. You can use the same options that we described earlier, such as choosing the color or line style of each curve if it's necessary.

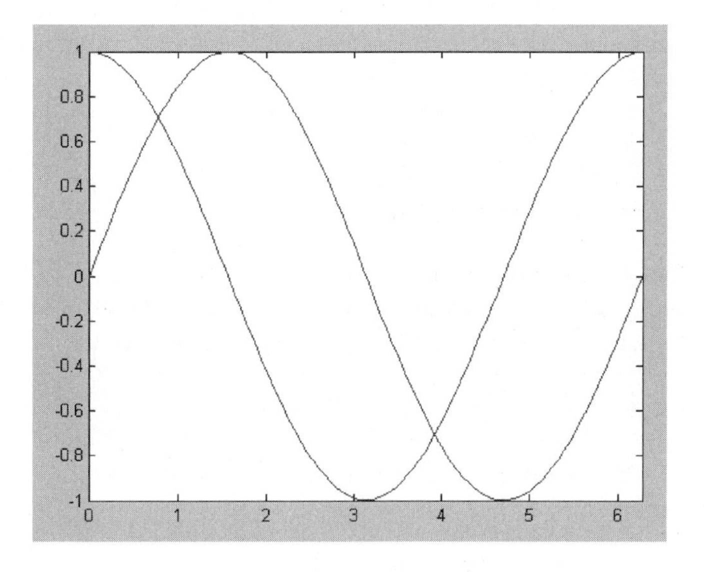

Figure 3-25 Both curves overlayed on the same graph

Polar and Logarithmic Plots

If you've taken calculus then no doubt you're familiar with polar and logarithmic plots. Back in my day we had to generate these manually—wouldn't it be nice to have a computer program do that for you or at least check your answers? Thankfully MATLAB comes to the rescue. Let's start by looking at polar plots, which plot the radius r as a function of polar angle θ.

For our first example let's generate a spiral. The so-called spiral of Archimedes is defined by the simple relationship:

$$r = a\theta$$

where a is some constant. Let's generate a polar plot of this function for the case where $a = 2$ and $0 \leq \theta \leq 2\pi$. The first statement we'll use defines the constant:

```
>> a = 2;
```

Well that was simple enough. Now let's define the function $r(\theta)$. This is done in two steps, first we have to treat θ the same way we would the independent variable x in our previous plots, so we define the label name, the range over which it is valid, and the increment we want to use. Next we define r:

```
>> theta = [0:pi/90:2*pi];
>> r = a*theta;
```

These statements tell MATLAB that *theta* is defined as $0 \leq \theta \leq 2\pi$. We have chosen our increment to be $\pi / 90$. The call to generate a polar plot is:

```
polar ( theta, r)
```

Let's call it and add a title to the plot:

```
>> polar(theta,r), title('Spiral of Archimedes')
```

The result is shown in Figure 3-26.

Many of the same options available with plot can be used with polar. As a second example, let's suppose that we want to generate a polar plot of the function:

$$r = 1 + 2\cos\theta$$

where $0 \leq \theta \leq 6\pi$, and display the resulting curve as a dashed line. First let's define our new range for θ

```
>> theta = [0:pi/90:6*pi];
```

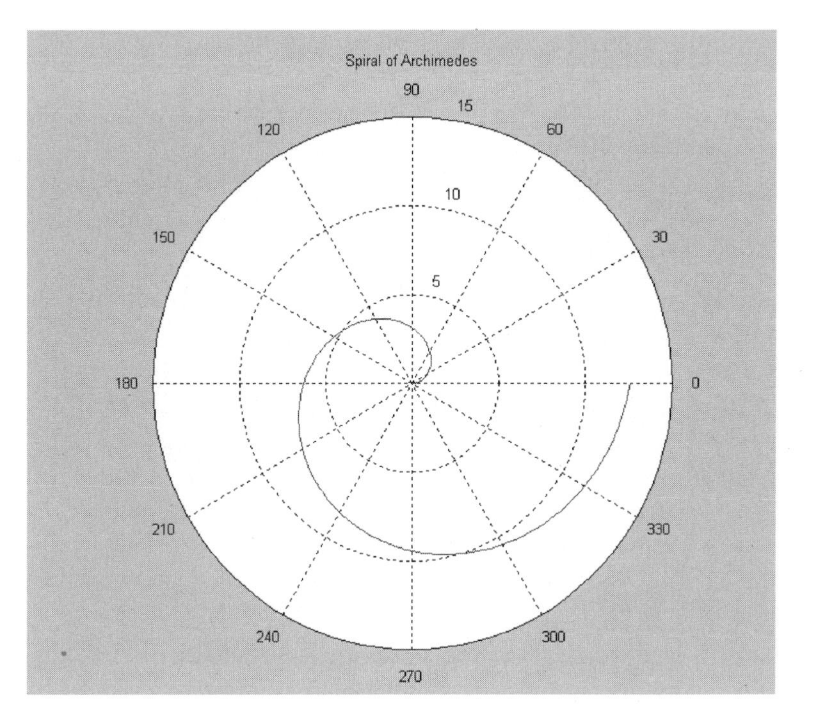

Figure 3-26 A polar plot of the spiral of Archimedes

Now we enter the function $r(\theta)$;

```
>> r = 1 + 2*cos(theta);
```

Let's tell MATLAB to draw a red curve with a dash-dot line. This is done by typing;

```
>> polar(theta,r,'r-.')
```

The plot that is generated is shown in Figure 3-27. Unfortunately in our black and white book you can't see the red line, but give it a try yourself to see the results.

Now let's take a look at how MATLAB can be used to generate logarithmic plots. This is something that used to give me headaches, and if you're an electrical engineer you will find this feature particularly useful. The first type of logarithmic plot we can use is the log-log plot. To see how this works, we are going to follow a typical example of an electrical circuit that consists of a voltage source, resistor, and capacitor. It is bound to be the case that many readers are not electrical engineers,

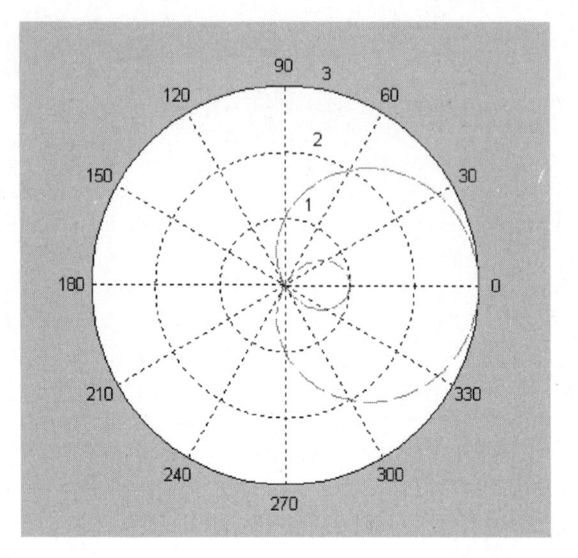

Figure 3-27 A polar plot of $r = 1 + 2\cos\theta$

so we aren't going to worry about the details of what's used to generate the equations, our purpose here is just to show you how to spit out a log-log plot.

It turns out that in an *RC* circuit that if the input voltage is sinusoidal where $v_i = A_i \sin \omega t$, then the output voltage will be some other sinusoidal function that we can write as $v_o = A_o \sin(\omega t + \phi)$. Something that is of interest to electrical engineers (and aren't they a strange lot) is the *frequency response.* This is the ratio of the output amplitude to the input amplitude, and for reasons that we can't understand this ratio turns out to be:

$$\frac{A_o}{A_i} = \left| \frac{1}{1 + i\omega RC} \right|$$

Basically the frequency response is going to tell us how strong the output signal is relative to the input signal at different frequencies. It's common to denote $s = i\omega$, since electrical engineers are so fond of the Laplace transform. So let's go ahead and do that, and let ω range over $1 \leq \omega \leq 100$ rad/s. The product of resistance and capacitance, RC has units of seconds. For our example, we will let $RC = 0.25$ seconds. Let's define these quantities in MATLAB:

```
>> RC = 0.25;
>> s = [1:100]*I;
```

Notice on the second line that we have defined s as a *complex* variable. The frequency response is a log-log plot of output/input ratio versus frequency. So let's define our function A_o/A_i. Notice from the definition we need the absolute value, this is done in MATLAB by passing the function to the abs command:

```
>> F = abs(1./(1+RC*s));
```

Now we've got everything we need to generate the plot. We can use many of the same options with loglog that can be applied to plot. First of all, chances are you want to include the grid on a log-log plot. We can do that and include axis labels and a graph title with the command:

```
>> loglog(imag(s),F),grid,xlabel('Frequency (rad/s)'),
ylabel('Output/Input Ratio'),title('Frequency Response')
```

This command generates the very nice log-log plot shown in Figure 3-28.

In my opinion the log-log plot was very easy to generate and we get nice output as a result. One way that log-log plots can be used is when a given function varies

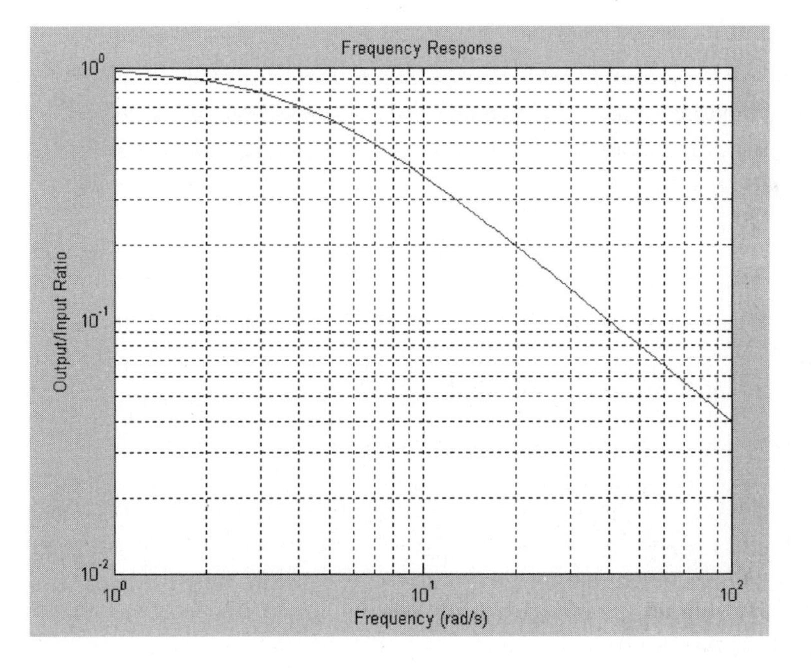

Figure 3-28 An example of a log-log plot generated using MATLAB

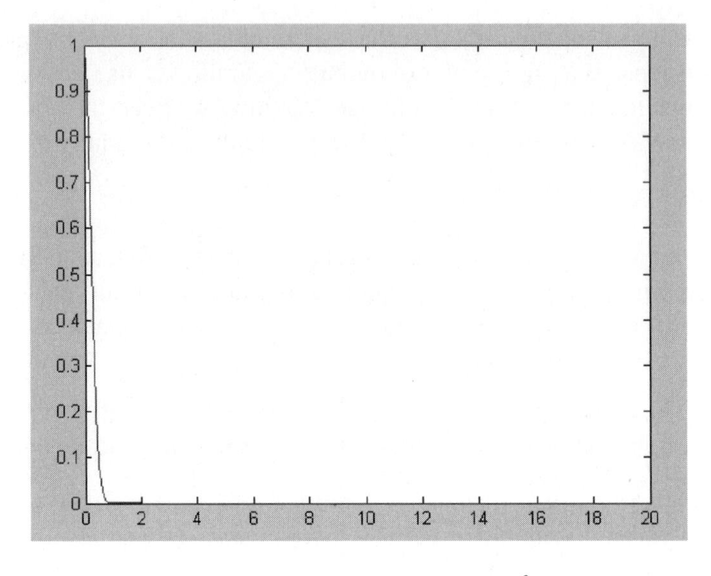

Figure 3-29 A plot of $y = e^{-10x^2}$

very rapidly over a small part of the domain. As a simple example, let's consider $y = e^{-10x^2}$, where $0 \leq x \leq 20$. We can generate a regular plot of this function with these commands:

```
>> x = [0:0.1:20];
>> y = exp(-10*x.^2);
>> plot(x,y);
```

As you can see in Figure 3-29, all the action occurs over a very tiny part of the data set.

Let's try a log-log plot of this function. We type:

```
>> loglog(x,y)
```

This generates the log plot in Figure 3-30. Notice that the close-up features of the function are present but we still have access to the entire range of data (with the exception of the origin where the log blows up to minus infinity).

We have two more options that are available, the first is *semilogx(x, y)* which generates a plot with a logarithmic x axis and a rectilinear y axis. Conversely, *semilogy*(x, y) generates a plot with a rectilinear x axis and a logarithmic y axis.

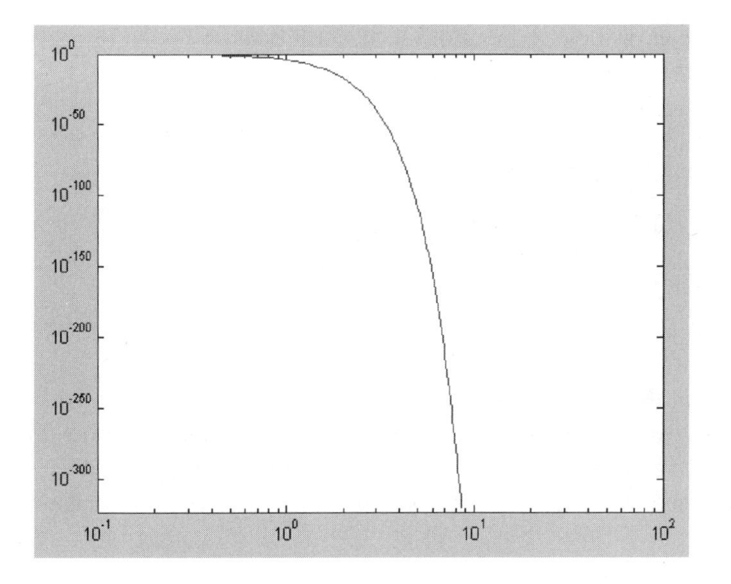

Figure 3-30 A log-log plot of $y = e^{-10x^2}$

Plotting Discrete Data

Now let's touch on some practical uses of MATLAB that you might need at work or in the field. First let's see how we can use plot(x, y) to plot a discrete set of data points and connect them with a line. Let's say we have a class with five students: Adrian, Jim, Joe, Sally, and Sue. They took an exam and received scores of 50, 98, 75, 80, and 98, respectively. How can we plot this data?

The first thing to do is define two arrays containing the list of students and the scores on the test. Our list of students plays the role of x in this case, so we just create an array of x values with 5 elements:

```
>> x = [1:5];
```

Since we aren't modeling a continuous function, it's not necessary to specify an increment. We are defaulting to the increment being 1, so MATLAB has just gone ahead and generated 5 points for us. Now let's put in the scores that correspond to each point, these are the y values, which we just create as a row vector. We enter the test scores as a comma delimited list with square brackets

```
>> y = [50,98,75,80,98];
```

Now we can plot the data, and tell MATLAB what labels to use. This is done with the *set* command. This looks complicated:

```
>> plot(x,y,'o',x,y),set(gca,'XTicklabel',['Adrian'; 'Jim';'
Joe';'Sally';'Sue']),...
set(gca,'XTick',[1:5]),axis([1 5 0 100]),xlabel('Student'),
ylabel('Final Test Score'),title('Final Exam December 2005')
```

Unfortunately, when we do this we get the following error:

```
??? Error using ==> vertcat
```

All rows in the bracketed expression must have the same number of columns.

Turns out we can't use the student names, each element in the list provided to set must have the same number of characters. So let's assign each student, a student ID number (something that would probably be done anyway). We make the following assignment:

```
'Adrian'; 'Jim';'Joe';'Sally';'Sue'] ‡  ['001';
'002';'003';'004';'005']
```

Now the command looks like:

```
plot(x,y,'o',x,y),set(gca,'XTicklabel',['001'; '002';'003';
'004';'005']), ...
set(gca,'XTick',[1:5]),axis([1 5 0 100]),xlabel('Student'),...
ylabel('Final Test Score'),title('Final Exam December 2005')
```

This generates the plot shown in Figure 3-31.

We can also plot the data as a two-dimensional bar chart. This is done by making a call to *bar*(*x, y*). This is a relatively simple and straightforward procedure. The commands to plot the test scores using bar(x, y) with axis labels and a chart title are:

```
>> x = [1:5];
>> y = [50,98,75,80,98];
>> bar(x,y), xlabel('Student'),ylabel('Score'),
title('Final Exam')
```

The result is shown in Figure 3-32.

The next way that is useful when plotting discrete data points is the *stem plot.* A stem plot generates a graph of a function with data at certain discrete points. At each point, a vertical line extends from the horizontal or x axis up to the value of the function at that point, which is marked off with a selected marker. As an example, let's consider the function $f(t) = e^{-\beta t}\sin(t/4)$, where $\beta = 0.01$ and pretend it represents the response of a spring to some force. Let's plot it in the continuous case to see

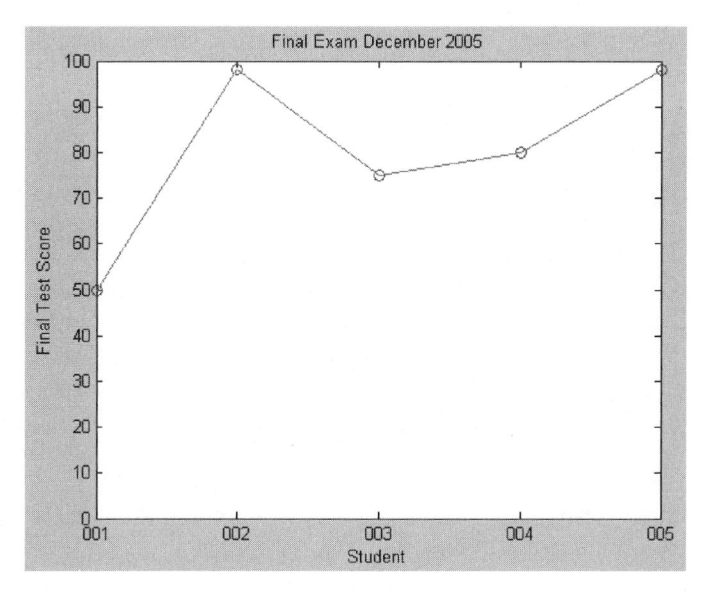

Figure 3-31 A plot of test scores with labels generated with the set command

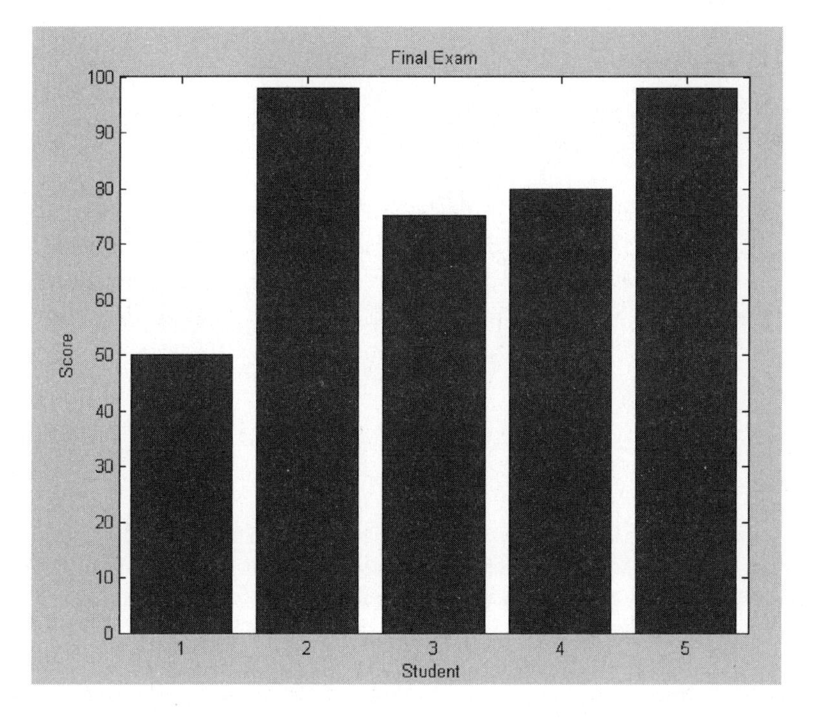

Figure 3-32 Test data shown on a bar chart

what it looks like when we take *t* out to 200 seconds. The commands we enter for the continuous plot are:

```
>>t = [0:0.1:200];
>> f = exp(-0.01*t).*sin(t/4);
>> plot(t,f),xlabel('time(sec)'),ylabel('spring response')
```

The plot is shown in Figure 3-33, showing the decaying oscillations of the spring.

Now let's suppose that the plot is going to be generated from a discrete data set. It might have been produced experimentally or in a computer simulation. We will suppose that the system was sampled every 5 seconds. We can pretend we've got this data by first generating a set of sampling times. We simply create our array of times with a time step of 5 seconds:

```
>> t = [0: 5: 200];
```

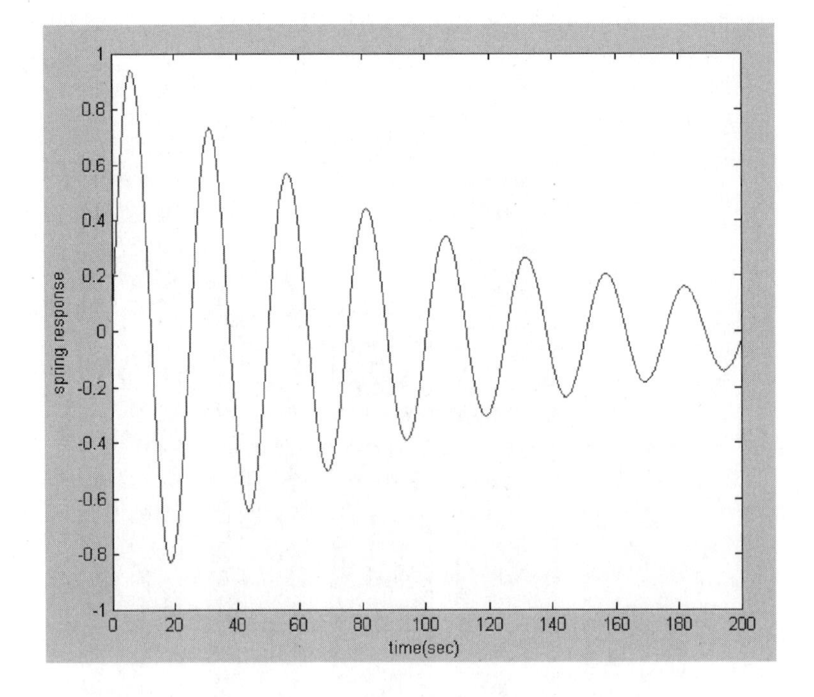

Figure 3-33 A plot of a decaying oscillation

If you had been doing an experiment or computer simulation, then you would have a set of output values that corresponded to each time sampled. To simulate this with known data, we simply define the function again and plot:

```
>> f = exp(-0.01*t).*sin(t/4)
>> plot(t,f),xlabel('time(sec)'),ylabel('spring response')
```

This time we get a somewhat crude representation of the function, shown in Figure 3-34.

In situations like this, engineers often like to generate stem plots. This is easily done in MATLAB using the *stem(x,y)* command. The command we use in this case is:

```
>> stem(t,f),xlabel('time(sec)'),ylabel('spring response')
```

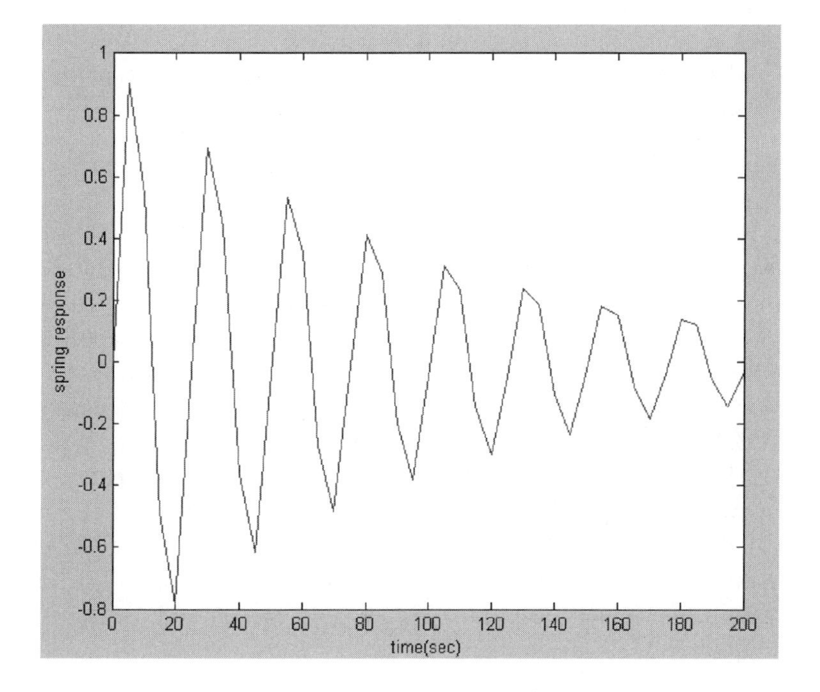

Figure 3-34 The same function shown in Figure 3-33, with a coarse sampling interval

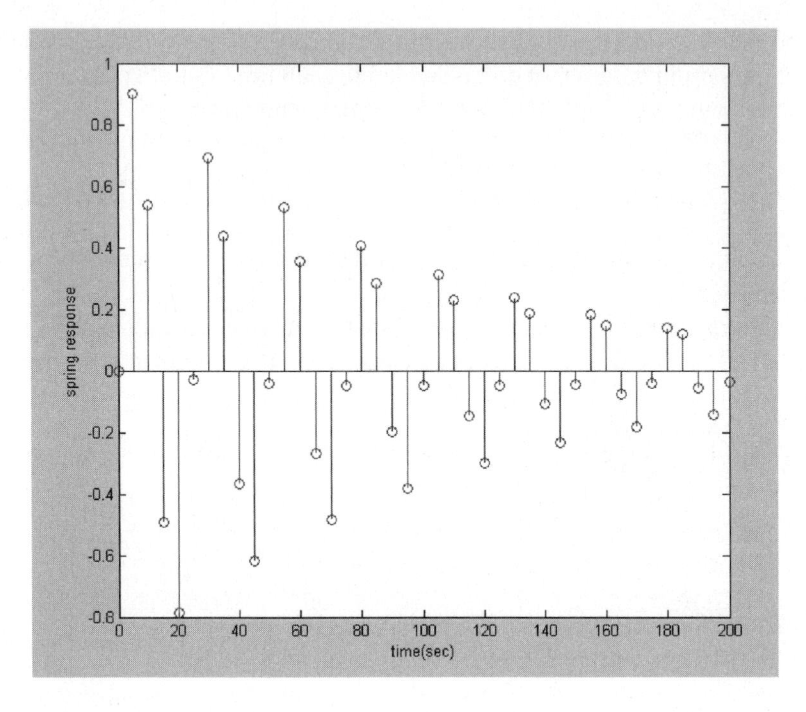

Figure 3-35 The spring response example generated using stem plot

The plot is shown in Figure 3-35.

The options available for line types using plot(x, y) can also be applied to stem. For example, we can have MATLAB generate dotted or dashed lines, or select which color we would like to use to draw the lines (blue is the default). We can also tell MATLAB to fill in the markers used at each data point by passing the character string 'fill' to stem(x,y). We also have the freedom to select the marker used, including square (s), diamond (d), five-pointed star (p), circle (o), cross (x), asterisk (*), and dot (.). Let's try filled diamonds, with green dashed lines. We tell MATLAB to use green dashed lines by adding the string '—g' to the arguments input to stem. Redrawing the plot in Figure 3-35 in this way produces the plot shown in Figure 3-36. The command is:

```
>> stem(t,f,'--dg','fill'),xlabel('time(sec)'),ylabel('spring response')
```

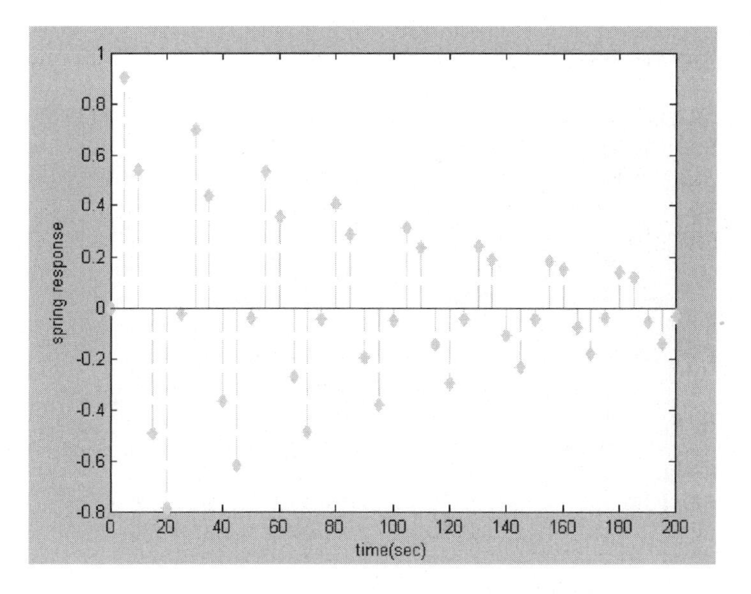

Figure 3-36 Stem plot generated using green, filled diamonds as markers and dashed lines

Contour Plots

Now let's see how MATLAB can be used to generate more sophisticated plots. We start by considering contour plots. The simplest type of contour we can generate is one that just shows the contour lines for a given function, but does not fill in the entire plot with colors. This is done by considering some function of two variables, $z = f(x, y)$. The first step, as always, is to generate the independent variables, which this time includes two sets of data x and y. This is done by calling the *meshgrid* command. Meshgrid is a straightforward extension of what we've already been doing to set up our independent variables. All it does is generate a matrix of elements that give the range over x and y we want to use along with the specification of increment in each case. So suppose that we're going to plot some function $z = f(x, y)$ where $-5 \le x \le 5$, $-3 \le y \le 3$. If we chose our increment in both cases to be 0.1, this simple call will do it:

```
>> [x,y] = meshgrid(-5:0.1:5,-3:0.1:3);
```

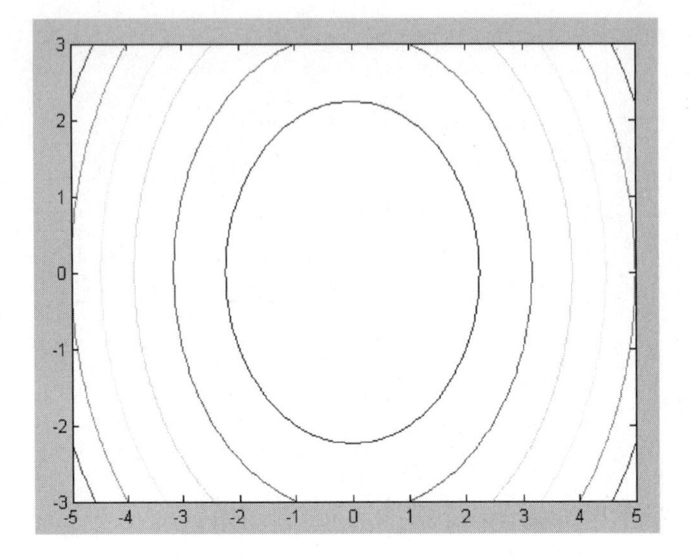

Figure 3-37 A contour plot of $z = x^2 + y^2$

Next we enter the function we want to use. For our first example, we try a simple function for which the contours are just circles, let's set $z = x^2 + y^2$. This is done by typing:

```
>> z = x.^2 + y.^2;
```

Now we call the contour command as follows:

```
>> contour(x,y,z)
```

This generates the plot shown in Figure 3-37.

OK that's a good start, but most contour plots you see have a bit more information than this. A good piece of info that is usually included is a label which tells you the constant value of each contour line. This can be done with a call to the set command. First we need to add a label we can use to reference the contour plot later, so we enter it this way:

```
[C,h] = contour(x,y,z);
```

Be sure to add the semicolon at the end so that the data isn't spit out at you. Now we call set:

```
>> set(h,'ShowText','on','TextStep',get(h,'LevelStep')*2)
```

This time, MATLAB puts labels on each curve, as shown in Figure 3-38.

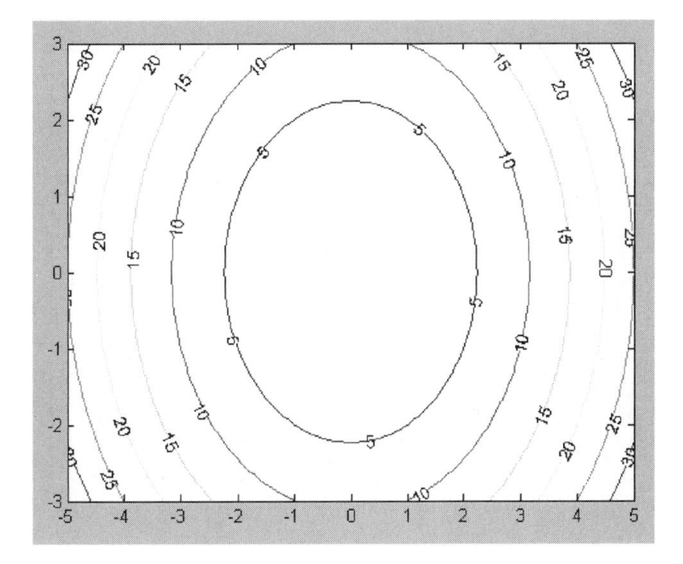

Figure 3-38 Adding labels to contour lines

Let's try the same thing with $z = \cos(x)\sin(y)$. The commands we use are:

```
>> z = cos(x).*sin(y);
>> [C,h] = contour(x,y,z);
>> set(h,'ShowText','on','TextStep',get(h,'LevelStep')*2)
```

This generates the contour map shown in Figure 3-39.

We can make the contour map three dimensional with a call to contour3. If we call contour3(z, n), it will generate a three-dimensional contour map with n contour levels. Let's stick with the same function $z = \cos(x)\sin(y)$. The call >> contour3(z,10) generates the plot shown in Figure 3-40.

This is helpful, but probably not as sophisticated as we would like. Let's make the plot more professional by telling MATLAB to fill in some more information. This time let's consider $z = ye^{-x^2+y^2}$, where $-2 \le x, y \le 2$. Since both variables have the same range of data, the meshgrid command is easier to enter:

```
>> [x,y] = meshgrid(-2:0.1:2);
>> z = y.*exp(-x.^2-y.^2);
```

That is, when both independent variables range over the same values, you can define the meshgrid as $[x, y] = \text{meshgrid}(x)$. Now let's do an ordinary contour plot of the function:

```
>> contour(x,y,z),xlabel('x'),ylabel('y')
```

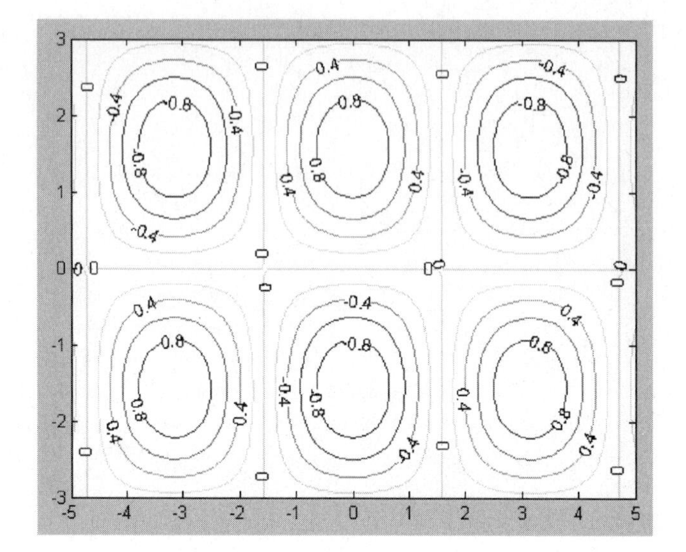

Figure 3-39 A contour map of $z = \cos(x)\sin(y)$ with $-5 \le x \le 5$, $-3 \le y \le 3$

The result is shown in Figure 3-41.

Next, we will generate the three-dimensional contour plot. If we just type in contour3(*x, y, z,* 30), we get the image shown in Figure 3-42.

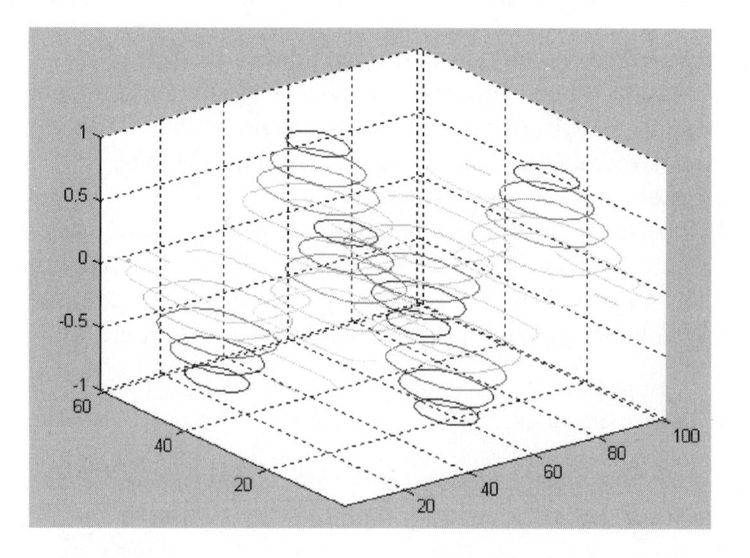

Figure 3-40 Our first call to contour3 for $z = \cos(x)\sin(y)$

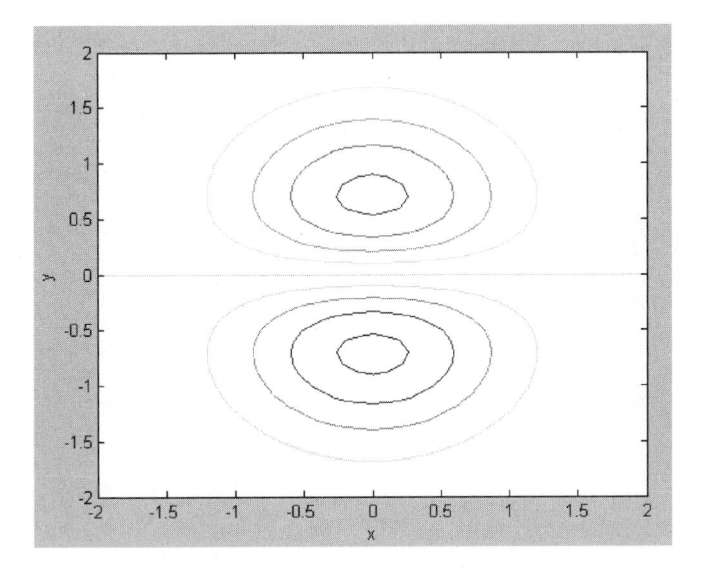

Figure 3-41 A contour plot of $z = ye^{-x^2+y^2}$

By adding the following surface command, we can spruce up the appearance of this plot quite a bit:

```
surface(x,y,z,'EdgeColor',[.8 .8 .8],'FaceColor','none')
grid off
view(-15,20)
```

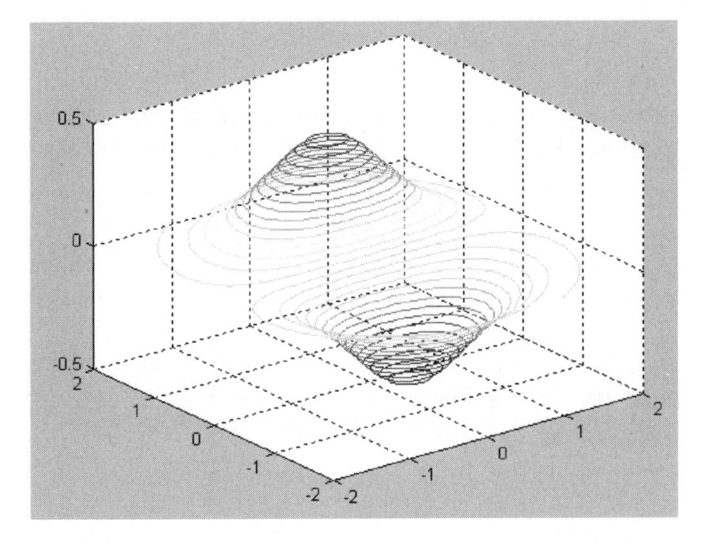

Figure 3-42 Three-dimensional contour plot of $z = ye^{-x^2+y^2}$

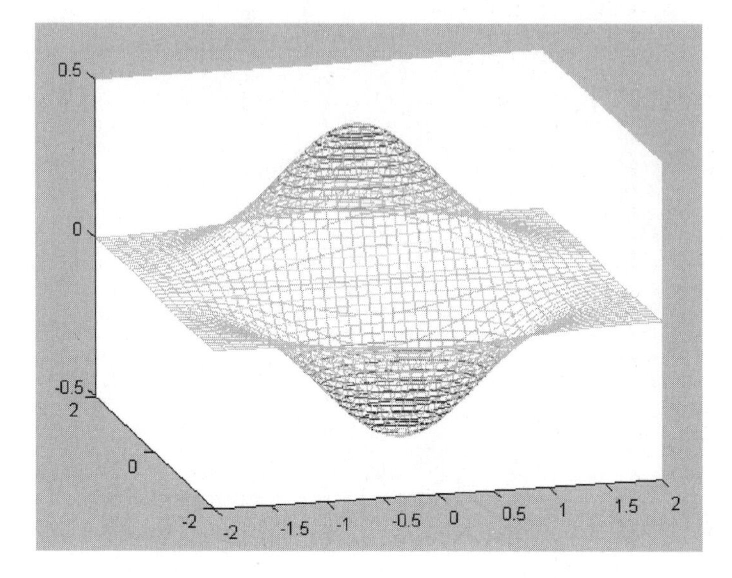

Figure 3-43 A dressing up of the contour plot using *surface* to show the surface plot of the function along with the contour lines

The result, shown in Figure 3-43, is reminiscent of the kinds of three-dimensional images you might see in your calculus book.

Three-Dimensional Plots

We have already seen a hint of the three-dimensional plotting capability using MATLAB when we considered the surface command. We can generate three-dimensional plots in MATLAB by calling *mesh*(x, y, z). First let's try this with $z = \cos(x)\sin(y)$ and $-2\pi \leq x, y \leq 2\pi$. We enter:

```
>> [x,y] = meshgrid(-2*pi:0.1:2*pi);
>> z = cos(x).*sin(y);
>> mesh(x,y,z),xlabel('x'),ylabel('y'),zlabel('z')
```

If you look at the final statement, you can see that mesh is just an extension of plot(x, y) to three dimensions. The result is shown in Figure 3-44.

Let's try it using $z = ye^{-x^2+y^2}$ with the same limits used in the last section. We enter the following commands:

```
>> [x,y] = meshgrid(-2:0.1:2);
>> z = y.*exp(-x.^2-y.^2);
>> mesh(x,y,z),xlabel('x'),ylabel('y'),zlabel('z')
```

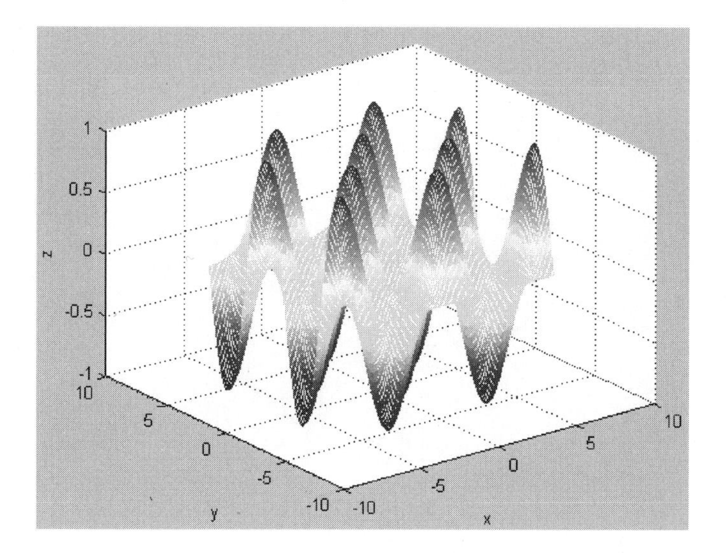

Figure 3-44 Plotting $z = \cos(x)\sin(y)$ using the mesh command

This plot is shown in Figure 3-45.

Now let's plot the function using a shaded surface plot. This is done by calling either *surf* or *surfc*. Simply changing the command used in the last example to:

```
>> surf(x,y,z),xlabel('x'),ylabel('y'),zlabel('z')
```

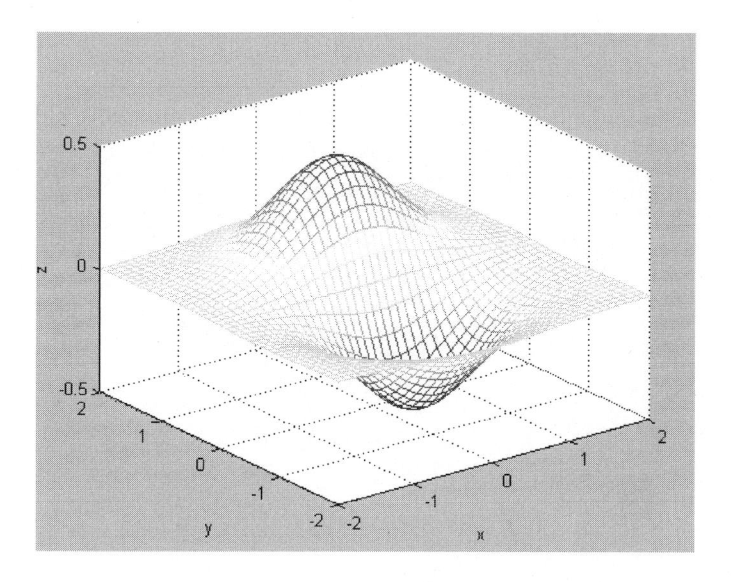

Figure 3-45 Plot of $z = ye^{-x_2 + y_2}$ generated with mesh

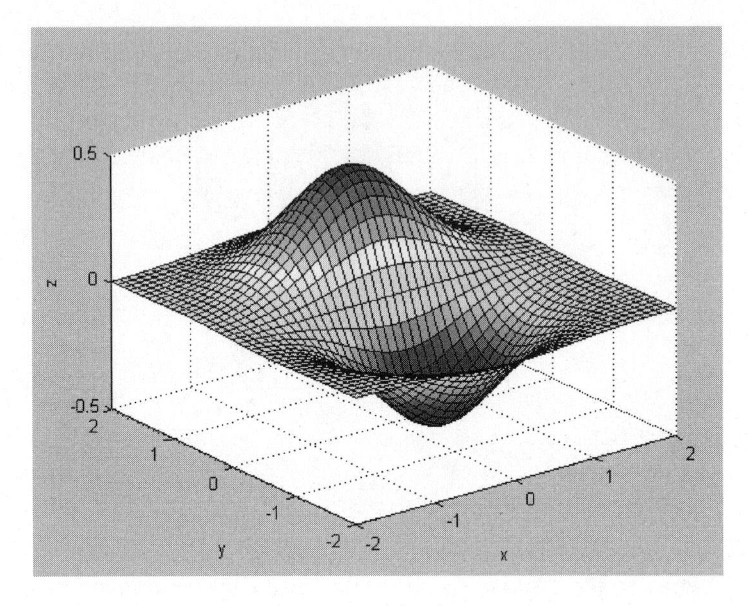

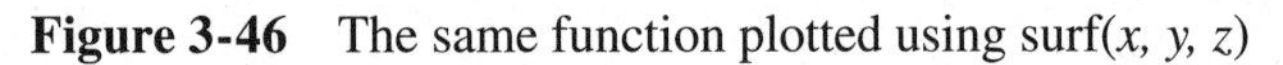

Figure 3-46 The same function plotted using surf(x, y, z)

Gives us the plot shown in Figure 3-46.

The colors used are proportional to the surface height at a given point. Using surfc instead includes a contour map in the plot, as shown in Figure 3-47.

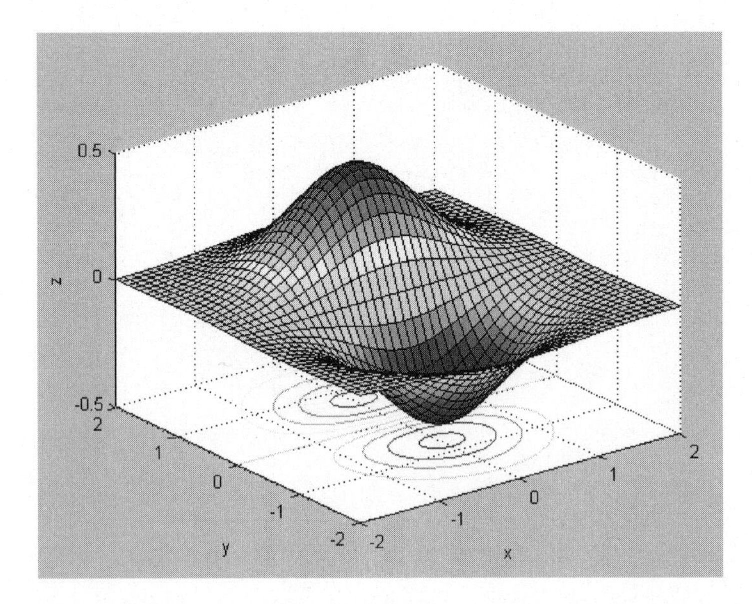

Figure 3-47 Using surfc displays contour lines at the bottom of the figure

Calling surfl (the ‘l’ tells us this is a lighted surface) is another nice option that gives the appearance of a three-dimensional illuminated object. Use this option if you would like a three-dimensional plot without the mesh lines shown in the other figures. Plots can be generated in color or grayscale. For instance, we use the following commands:

```
>> surfl(x,y,z),xlabel('x'),ylabel('y'),zlabel('z')
>> shading interp
>> color map(gray);
```

This results in the impressive grayscale plot of $z = ye^{-x^2+y^2}$ shown in Figure 3-48.

The *shading* used in a plot can be set to flat, interp, or faceted. Flat shading assigns a constant color value over a mesh region with hidden mesh lines, while faceted shading adds the meshlines. Using interp tells MATLAB to interpolate what the color value should be at each point so that a continuously varying color map or grayscale shading scheme is generated, as we considered in Figure 3-48. Let’s show the three variations using color (well unfortunately you can’t see it, but you can try it).

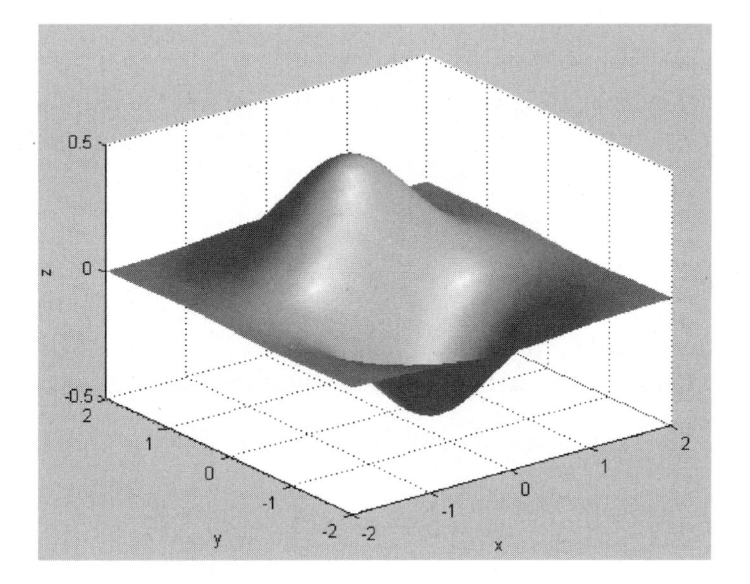

Figure 3-48 A plot of the function using surfl

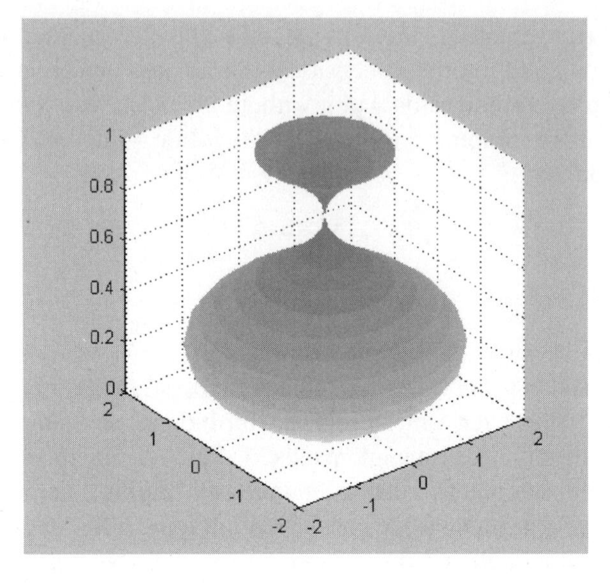

Figure 3-49 Plot generated using the cylinder function and flat shading

Let's generate a funky cylindrical plot. You can generate plots of basic shapes like spheres or cylinders using built-in MATLAB functions. In our case, let's try

```
>> t = 0:pi/10:2*pi;
[X,Y,Z] = cylinder(1+sin(t));
surf(X,Y,Z)
axis square
```

If we set shading to flat, we get the funky picture shown in Figure 3-49 that looks kind of like a psychedelic raindrop.

Now let's try a slightly different function, this time going with faceted shading. We enter:

```
>> t = 0:pi/10:2*pi;
[X,Y,Z] = cylinder(1+cos(t));
surf(X,Y,Z)
axis square
>> shading faceted
```

This gives us the plot shown in Figure 3-50.

If we go to shading interp, we get the image shown in Figure 3-51.

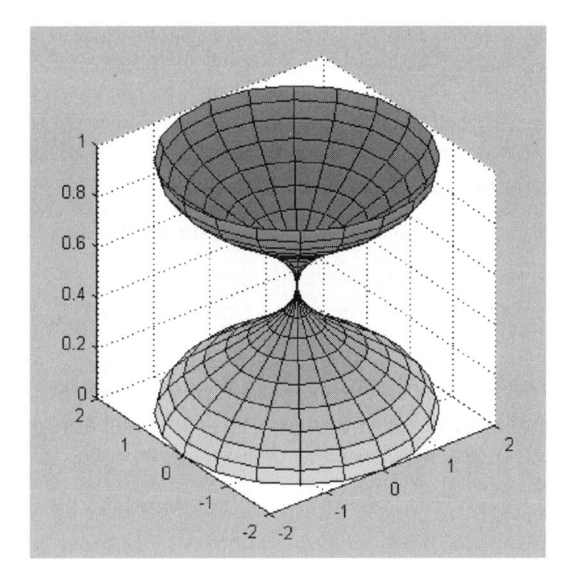

Figure 3-50 Cylindrical plot of cylinder[1+cos(t)] using faceted shading

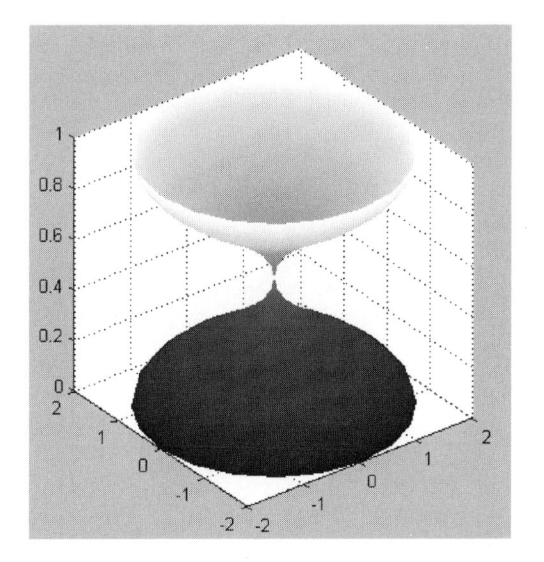

Figure 3-51 Plot of cylinder[1+cos(t)] with interpolated shading

Quiz

1. Generate a plot of the tangent function over $0 \le x \le 1$ labeling the x and y axes. Set the increment to 0.1.
2. Show the same plot, with sin(x) added to the graph as a second curve.
3. Create a row vector of data points to represent $-\pi \le x \le \pi$ with an increment of 0.2. Represent the same line using linspace with 100 points and with 50 points.
4. Generate a grid for a three-dimensional plot where $-3 \le x \le 2$ and $-5 \le y \le 5$. Use an increment of 0.1. Do the same if $-5 \le x \le 5$ and $-5 \le y \le 5$ with an increment of 0.2 in both cases.
5. Plot the curve $x = e^{-t} \cos t$, $y = e^{-t} \sin t$, $z = t$ using the plot3 function. Don't label the axes, but turn on the grid.

CHAPTER 4

Statistics and an Introduction to Programming in MATLAB

The ease of use available with MATLAB makes it well suited for handling probability and statistics. In this chapter we will see how to use MATLAB to do basic statistics on data, calculate probabilities, and present results. To introduce MATLAB's programming facilities, we will develop some examples that solve problems using coding and compare them to solutions based on built-in MATLAB functions.

Generating Histograms

At the most basic level, statistics involves a discrete set of data that can be characterized by a mean, variance, and standard deviation. The data might be plotted in a bar chart. We will see how to accomplish these basic tasks in MATLAB in the context of a simple example.

Imagine a ninth grade algebra classroom that has 36 students. The scores received by the students on the midterm exam and the number of students that obtained each score are:

One student scored 100

Two students scored 96

Four students scored 90

Two students scored 88

Three students scored 85

One student scored 84

Two students scored 82

Seven students scored 78

Four students scored 75

Six students scored 70

One student scored 69

Two students scored 63

One student scored 55

The first thing we can do in MATLAB is enter the data and generate a bar chart or histogram from it. First, we enter the scores (x) and the number of students receiving each score(y):

```
>> x = [55, 63,69,70,75,78,82,84,85,88,90,96,100];
>> y = [1,2,1,6,4,7,2,1,3,2,4,2,1];
```

We can generate a bar chart with a simple call to the *bar* command. It works just like plot, we just pass the x data and the y data in two arrays with the call bar (x, y). For instance, we can quickly generate a histogram with the data we've got:

```
>> bar(x,y)
```

This generates the bar chart shown in Figure 4-1. But this isn't really satisfying.

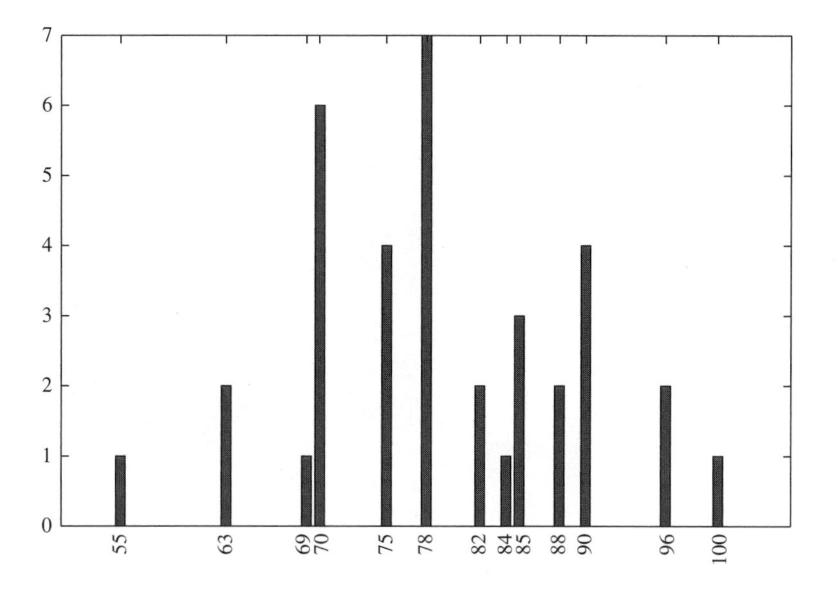

Figure 4-1 Our first stab at generating a bar chart of exam scores

If you're the teacher what you really probably want to know is how many students got As, Bs, Cs, and so on. One way to do this is to manually collect the data into the following ranges:

One student scored 50–59

Three students scored 60–69

Seventeen students scored 70–79

Eight students scored 80–89

Seven students scored 90–100

So we would generate two arrays as follows. First we give MATLAB the midpoint of each range we want to include on the bar chart. The midpoints in our case are:

```
>> a = [54.5,64.5,74.5,84.5,94.5];
```

Next we enter the number of students that fall in each range:

```
>> b = [1,3,17,8,7];
```

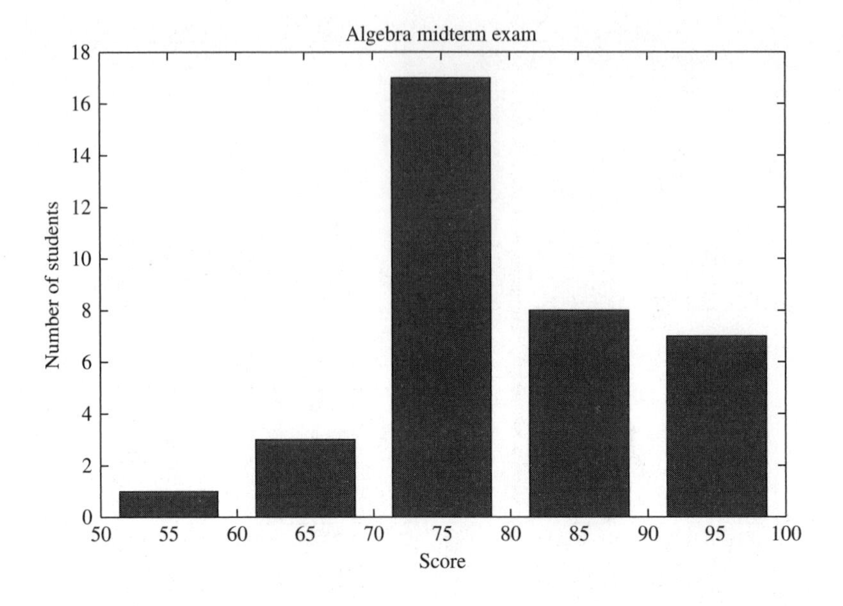

Figure 4-2 Making a nicer bar chart by binning the data

Now we call bar(*x*, *y*) again, adding axis labels and a title:

```
>> bar(a,b),xlabel('Score'),ylabel('Number of Students'),
title('Algebra Midterm Exam')
```

The result, shown in Figure 4-2, is that we have produced a more professional looking bar chart of the data.

While MATLAB has a built-in histogram function called *hist*, you may find it producing useful charts on a part-time basis. Your best bet is to manually generate a bar chart the way we have done here. Before moving on we note some variations on bar charts you can use to present data. For example, we can use the *barh* command to present a horizontal bar chart:

```
>> barh(a,b),xlabel('Number of Students'),ylabel('Exam
Score')
```

The result is shown in Figure 4-3.

You might also try a fancy three-dimensional approach by calling bar3 or bar3h. This can be done with the following command and is illustrated in Figure 4-4:

```
>> bar3(a,b),xlabel('Exam Score'),ylabel('Number of Students')
```

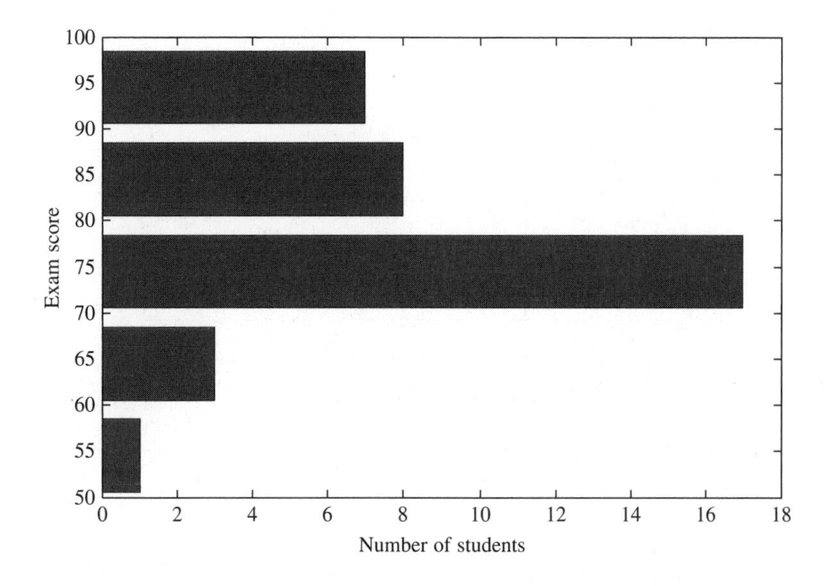

Figure 4-3 Presenting the exam data using a horizontal bar chart

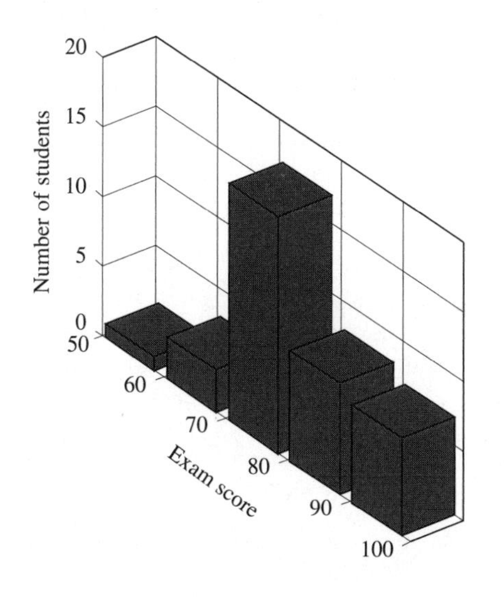

Figure 4-4 A three-dimensional bar chart

EXAMPLE 4-1

Three algebra classes at Central High taught by Garcia, Simpson, and Smith take their midterm exams with scores:

Score Range	Garcia	Simpson	Smith
90–100	10	5	8
80–89	13	10	17
70–79	18	20	15
60–69	3	5	2
50–59	0	3	1

Generate grouped and stacked histograms of the data.

SOLUTION 4-1

Bar charts created in MATLAB with multiple data sets can be grouped or stacked together. To generate a grouped bar chart of *x, y* data, we write bar(x, y, grouped′). The default option is grouped, so bar(x, y) will produce exactly the same result. To generate a stacked bar chart, we write bar(x, y, stacked′). The data is entered by creating a multiple column array, with one column representing the scores for Garcia's class, the next column representing the scores for Simpson's class, and the final column representing the scores for Smith's class. First we create the *x* array of data, which contains the midpoint of the score ranges:

```
>> x = [54.5,64.5,74.5,84.5,94.5];
```

Now let's enter the scores in three column vectors:

```
>> garcia = [0; 3; 18; 13; 10]; simpson = [3; 5; 20; 10; 5];
smith = [1; 2; 15; 17; 8]
```

The next step that is required is to put all the data in a single array. We can create a single array that has columns corresponding to the garcia, simpson, and smith arrays with this command:

```
>> y = [garcia simpson smith];
```

Now we can generate a histogram:

```
>> bar(x,y),xlabel('Exam Score'),ylabel('Number of Students')
,legend('Garcia','Simpson','Smith')
```

The result is shown in Figure 4-5.

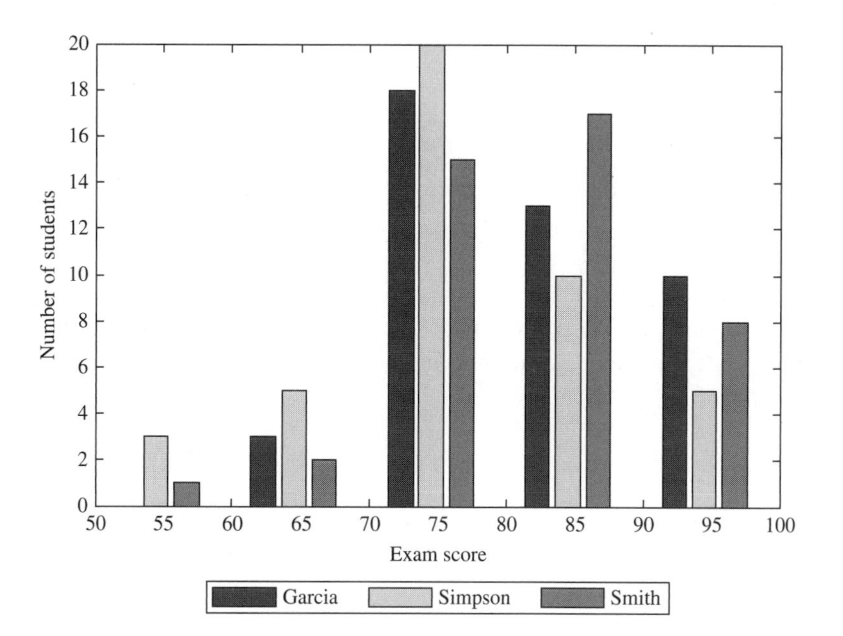

Figure 4-5 A grouped bar chart

Basic Statistics

MATLAB can tell us what the mean of a set of data is with a call to the *mean* function:

```
>> a = [11,12,16,23,24,29];
>> mean(a)

ans =

   19.1667
```

We can pass an array to mean and MATLAB will tell us the mean of each column:

```
>> A = [1 2 3; 4 4 2; 4 2 9]

A =
```

```
     1     2     3
     4     4     2
     4     2     9

>> mean(A)

ans =
    3.0000    2.6667    4.6667
```

But this simple built-in function won't work for when the data is weighted. We'll need to calculate the mean manually. The mean of a set of data x_j is given by:

$$\langle x_j \rangle = \frac{\sum_{j=1}^{N} x_j N(x_j)}{\sum_{j=1}^{N} N(x_j)}$$

where $N(x_j)$ is the number of elements with value x_j. Let's clarify this with an example—we return to our original set of test scores.

To see how to do basic statistics on a set of data we return to our first example. We have a set of test scores (x) and the number of students receiving each score(y):

```
>> x = [55, 63,69,70,75,78,82,84,85,88,90,96,100];
>> y = [1,2,1,6,4,7,2,1,3,2,4,2,1];
```

In this case the x array represents x_j while the y array represents $N(x_j)$. To generate the sum:

$$\sum_{j=1}^{N} N(x_j)$$

We simply sum up the elements in the y array:

```
>> N = sum(y)

N =

    36
```

Now we need to generate the sum:

$$\sum_{j=1}^{N} x_j N(x_j)$$

This is just the vector multiplication of the x and y arrays. Let's denote this intermediate sum by s:

```
>> s = sum(x.*y)

s =

        2847
```

Then the average score is:

```
>> ave = s/N

ave =

   79.0833
```

With the data entered in two arrays, we can find the probability of obtaining different results. The probability of obtaining result x_j is:

$$P(x_j) = \frac{N(x_j)}{N}$$

For example, the probability that a student scored 78, which is the sixth element in our array is:

```
>> p = y(6)/N

p =

    0.1944
```

We can use this definition of probability to rewrite the average as:

$$\langle x_j \rangle = \frac{\sum_{j=1}^{N} x_j N(x_j)}{\sum_{j=1}^{N} N(x_j)} = \sum_{i=1}^{N} x_j p(x_j)$$

First let's generate an array of probabilities:

```
>> p = y/N

p =

  Columns 1 through 10

    0.0278    0.0556    0.0278    0.1667    0.1111    0.1944
0.0556    0.0278    0.0833    0.0556

  Columns 11 through 13

    0.1111    0.0556    0.0278
```

Then the average could be calculated using:

```
>> ave = sum(x.*p)

ave =

   79.0833
```

Writing Functions in MATLAB

Since we're dealing with the calculation of some basic quantities using sums, this is a good place to introduce some basic MATLAB programming. Let's write a routine (a *function*) that will compute the average of a set of weighted data. We'll do this using the formula:

$$\frac{\sum_{j=1}^{N} x_j N(x_j)}{\sum_{j=1}^{N} N(x_j)}$$

To create a function that we can call in the command window, the first step is to create a .m file. To open the file editor, we use these two steps:

1. Click the File pull-down menu
2. Select New → m file

This opens the file editor that you can use to type in your script file. Line numbers are provided on the left side of the window. On line 1, we need to type in the word *function,* along with the name of the variable we will use to return the data, the function name, and any arguments we will pass to it. Let's call our function *myaverage.* The function will take two arguments:

- An array x of data values
- An array N that contains the number at each data value $N(x)$

We'll use a function variable called ave to return the data. The first line of code looks like this:

```
function ave = myaverage(x,N)
```

To compute the average correctly, x and N must contain the same number of elements. We can use the size command to determine how many elements are in each array. We will store the results in two variables called size$x$ and size$N$:

```
sizex = size(x);
sizeN = size(N);
```

The variables size$x$ and size$N$ are actually row vectors with two elements. For example if x has four data points then:

```
sizex =

     1     4
```

So to test the values to see if they are equal, we will have to check size$x$(2) and size$N$(2). One way to test them with an if statement is to ask if size$x$ is greater than size$N$ OR size$N$ is greater than size$x$. We indicate OR in MATLAB with a "pipe" character, i.e. |. So this would be a valid way to check this condition:

```
if (sizex(2) > sizeN(2)) | (sizex(2) <sizeN(2))
```

Another way is to simply ask if size$x$ and size$N$ are not equal. We indicate not equal by preceding the equal sign with a tilde, in other words if size$x$ is NOT EQUAL to size$N$ would be implemented by writing:

```
if sizex(2) ~= sizeN(2)
```

If the two sizes are not equal, then we want to terminate the function. If they are equal, we will go ahead and calculate the average. This can be implemented using an *if –else* statement. What we will do is use the disp command to print an error message to the screen if the user has passed two arrays that are different sizes. The first part of the if-else statement looks like this:

```
if sizex(2) ~= sizeN(2)
 disp('Error: Arrays must be same dimensions')
```

Notice that we have used the operator "~=" to represent *not equal* in MATLAB. In the else part of the statement, we add statements to calculate the average. First let's sum up the number of sample points. This is:

$$\sum_{j=1}^{N} N(x_j)$$

We can do this in MATLAB by calling the sum command, passing N as the argument. So the next two lines in the function are:

```
else
    total = sum(N);
```

Now we add a line to compute the numerator in the average formula:

$$\sum_{j=1}^{N} x_j N(x_j)$$

This is the line we need:

```
s = x.*N;
```

Finally, we compute the average and assign it to the variable name we used in the function declaration:

```
ave = sum(s)/total;
```

An end statement is used to terminate the if-else block. The completed function looks like this:

```
function ave = myaverage(x,N)
sizex = size(x);
sizeN = size(N);
if sizex(2) ~= sizeN(2)
    disp('Error: Arrays must be same dimensions')
else
    total = sum(N);
```

```
    s = x.*N;
    ave = sum(s)/total;
end
```

Once the function is written, we need to save it so that it can be used in the command window. MATLAB will save your .m files in the *work* folder.

Let's return to the command window and see how we can use the function. At an area law office, there are employees with the following ages:

Age	Number of Employees
20	2
25	3
38	4
43	2
55	3

Let's create the age array, which corresponds to x in our function:

```
>> age = [20, 25, 38, 43, 55];
```

Next we create an array called num which corresponds to N in our function:

```
>> num = [2, 3, 4, 2, 3];
```

Calling the function, we find the average age is:

```
ans =

    37
```

Before moving on let's test our function to make sure it reports the error message when we pass two arrays that aren't the same size. Let's try:

```
>> a = [1,2,3];
>> b = [1,2,3,4,5,6];
```

When we call myaverage we get the error message:

```
>> myaverage(a,b)
```

Error: Arrays must be same dimensions

Programming with For Loops

A *For Loop* is an instruction to MATLAB telling it to execute the enclosed statements a certain number of times. The syntax used in a For Loop is:

```
for index = start: increment : finish
   statements
end
```

We can illustrate this idea by writing a simple function that sums the elements in a row or column vector. If we leave the increment parameter out of the For Loop statement, MATLAB assumes that we want the increment to be one.

The first step in our function is to declare the function name and get the size of the array passed to our function:

```
function sumx = mysum(x)
%get number of elements
num = size(x);
```

We've added a new programming element here—we included a comment. A comment is an explanatory line for the reader that is ignored by MATLAB. We indicate comments by placing a % character in the first column of the line. Now let's create a variable to hold the sum and initialize it to zero:

```
%initialize total
sumx = 0;
```

Now we use a For Loop to step through the elements in the column vector 1 at a time:

```
for i = 1:num(2)
 sumx = sumx + x(i);
end
```

When writing functions, be sure to add a semicolon at the end of each statement—unless you want the result to be displayed on screen.

Calculating Standard Deviation and Median

Let's return to using the basic statistical analysis tools in MATLAB to find the standard deviation and median of a set of discrete data. We assume the data is given in terms of a frequency or number of data points. Once again, as a simple example we consider a set of employees at an office, and we are given the number of employees at each age. Suppose that there are:

Two employees aged 17

One employee aged 18

Three employees aged 21

One employee aged 24

One employee aged 26

Four employees aged 28

Two employees aged 31

One employee aged 33

Two employees aged 34

Three employees aged 37

One employee aged 39

Two employees aged 40

Three employees aged 43

The first thing we will do is create an array of absolute frequency data. This is the array $N(j)$ that we had been using in the previous sections. This time we will have an entry for each age, so we put a 0 if no employees are listed with the given age. Let's call it f_abs for absolute frequency:

```
f_abs = [2, 1, 0, 0, 3, 0, 0, 1, 0, 1, 0, 4, 0, 0, 2, 0, 1,
2, 0, 0, 3, 0, 1, 2, 0, 0, 3];
```

We are "binning" the data, so let's define a bin width. Since we are measuring this in one year increments, we set our bin width to one:

```
binwidth = 1;
```

Now we create an array that represents the ages ranged from 17 to 43 with a binwidth of one year:

```
bins = [17:bin width:43];
```

Now we collect the raw data. We use a For Loop to sweep through the data as follows:

```
raw = [];
for i = 1:length(f_abs)
if f_abs(i) > 0
  new = bins(i)*ones(1,f_abs(i));
```

```
else
  new = [];
end
raw =[raw,new];
end
```

What this loop does is it creates an array with individual elements repeated by frequency:

```
raw =

  Columns 1 through 18

    17    17    18    21    21    21    24    26    28    28
28    28    31    31    33    34    34    37

  Columns 19 through 26

    37    37    39    40    40    43    43    43
```

So we have replaced the approach we used in the last section of having two arrays, one with the given age and one with the frequency at that age, by a single array that has each age repeated the given number of times. We can use this raw data array with the built-in MATLAB functions to compute statistical information. For example, the mean or average age of the employees is:

```
>> ave = mean(raw)

ave =

   30.7308
```

We might also be interested in the median. This tells us the age for which half the employees are younger, and half are older:

```
>> md = median(raw)

md =

    31
```

The standard deviation is:

```
>> sigma = std(raw)

sigma =

    8.3836
```

If the standard deviation is small, that means most of the data is near the mean value. If it is large, then the data is more scattered. Since our bin size is 1 year in this case, an 8.4 year standard deviation indicates the latter situation applies to this data. Let's plot a scaled frequency bar chart to look at the shape of the data. The first step is to calculate the "area" of the data:

```
area = binwidth*sum(f_abs);
```

Now we scale it:

```
scaled_data = f_abs/area;
```

And generate a plot:

```
bar(bins,scaled_data),xlabel('Age'),ylabel('Scaled Frequency')
```

The result is shown in Figure 4-6.

As you can see from the plot, the data doesn't fit neatly around the mean, which is about 31 years. For a contrast let's consider another office with a nice distribution. Suppose that the employees range in age 17–34 with the following frequency distribution:

```
f_abs = [2, 1, 0, 0, 5, 4, 6, 7, 8, 6, 4, 3, 2, 2, 1, 0, 0, 1];
```

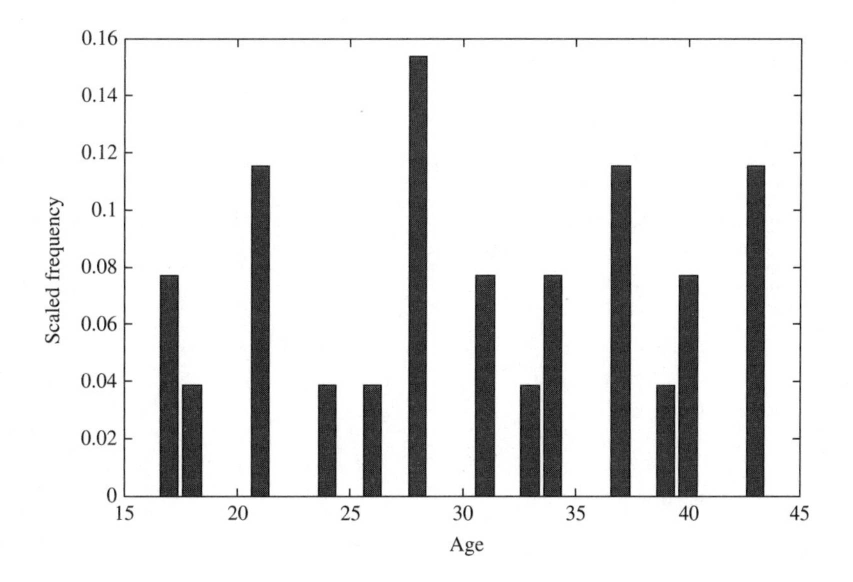

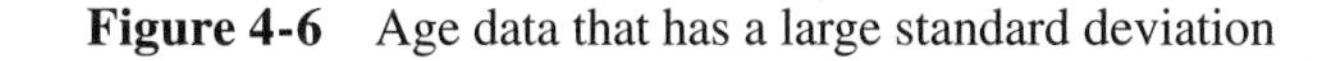

Figure 4-6 Age data that has a large standard deviation

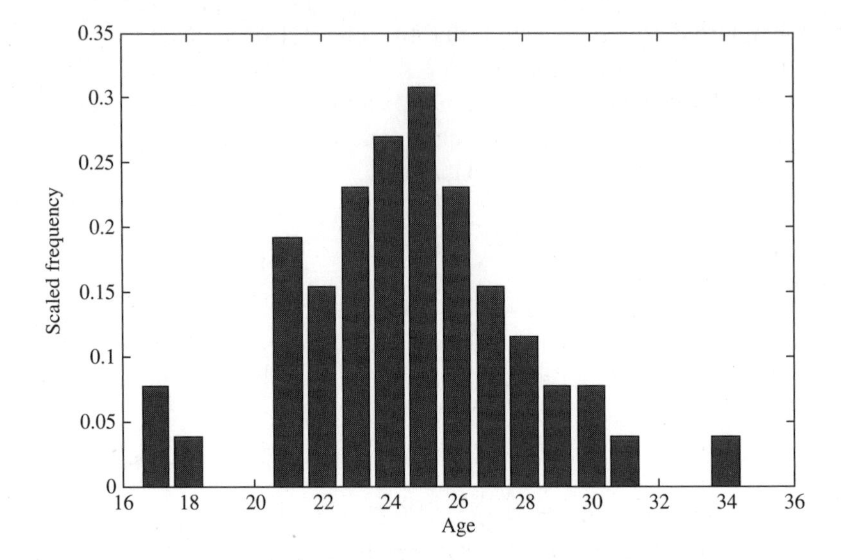

Figure 4-7 A data set with a smaller standard deviation

If we go through the same process with this data, we get the scaled frequency bar chart shown in Figure 4-7.

The basic statistical data for this set of employees is:

```
>> mu = mean(raw)

mu =

   24.6538

>> med = median(raw)

med =

    25

>> sigma = std(raw)

sigma =

    3.3307
```

The standard deviation is much smaller—and we can see it in the scaled frequency plot. It is approaching the shape of a Gaussian or bell curve, which we show in Figure 4-8.

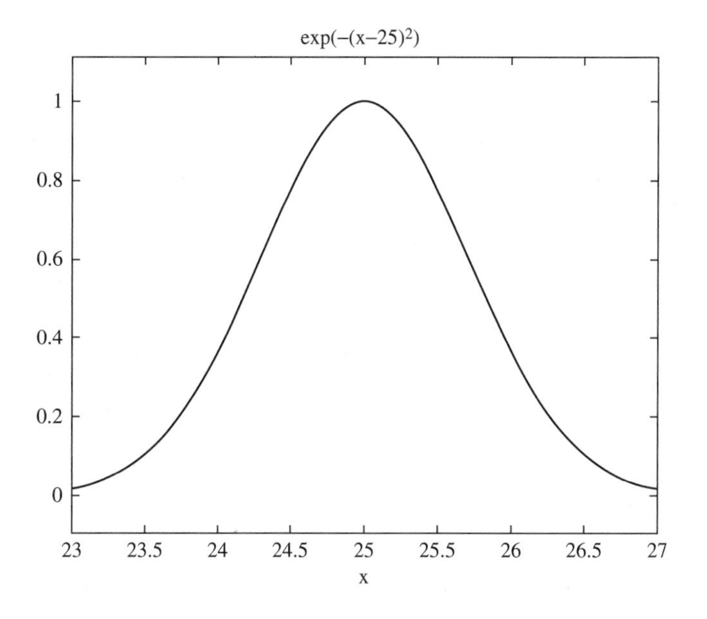

Figure 4-8 A bell curve with a mean of 25 years

When the data fits a Gaussian distribution, the standard deviation can be used to characterize the data to determine the probability of it falling at a certain value. Let's just treat the last data set as if this were the case. The standard deviation is denoted σ while the mean is denoted by μ. Then the percentage of the graph that lies in the ranges:

$$\mu - \sigma \leq x < \mu + \sigma,\ \mu - 2\sigma \leq x < \mu + 2\sigma, \text{ and } \mu - 3\sigma \leq x \leq \mu + 3\sigma$$

is 68%, 96%, and 99.7%, respectively. Using the previous set of age data, the standard deviation and mean were found to be:

```
sigma =

    3.3307

>> mu

mu =

   24.6538
```

About 68% of the ages will be in 1 standard deviation of the mean, that is, between mu – sigma = 21.3232 years and mu + sigma = 27.9845 years. Going further, 96% of the ages will be between mu – 2*sigma = 17.9925 years and mu + 2*sigma = 31.3152 years. The correspondence is not exact, because our data set is not exactly a bell curve, but we are illustrating the idea and how you could use MATLAB to analyze data that is close to a bell curve.

More MATLAB Programming Tips

Earlier we wrote a script file (.m) to implement a function which could be written up like computer code. Let's digress a bit and round out the chapter looking at a few more programming structures. First let's see how to get data from the user or from yourself in real time at a computer prompt. This is done using the *input* function. To use *input,* you set a variable you want to use to store the input data equal to the command, which takes a quote-delimited string as argument. When the statement is executed MATLAB prints your string to the command window and waits for data to be typed in. When the return key is hit, it assigns whatever data the user typed to the appropriate variable.

Let's concoct a simple example to see how this works. Suppose that we want to get the number of square feet of a house, and given that the price of homes in Podunk, Maine is $10 per square foot, we output the total asking price of the house. We can ask for and get the data using the following commands.

First, since we are working with financial data here, we tell MATLAB to display results with the bank format. Then we set our rate per square foot.

```
>> format bank
>> rate = 10;
```

Now let's ask the user to enter the square footage of the house by politely asking for it with the input command:

```
>> sqft = input('Enter total sqft of house: ')
```

MATLAB responds thus:

```
Enter total sqft of house:
```

If you enter these statements, MATLAB will be waiting patiently with a blinking cursor just to the right of this statement. Let's say we enter 1740. The result is:

```
sqft =

        1740.00
```

Since we have selected format bank, MATLAB is using two decimal places in its calculation. Unfortunately, it does not appear polite enough to print a dollar sign for us.

Now let's do our calculation:

```
>> price = rate*sqft

price =

      17400.00
```

Now real estate is very cheap in Podunk, Maine, but that is not our concern. How do we get this info back out to the world? We do it using the disp command. Let's tell the user what we are spitting out and then print our result:

```
>> disp('The total price is $'), disp(price)
The total price is $:
      17400.00
```

Alright at this point we have learned two important things—how to get data from the user and how to return answers to the user. Now let's see how to start controlling program flow.

Here is another example. We write a function to ask the user for the radius of a sphere. We use it to calculate the volume of the sphere, and output the result to the user:

```
function volume
%ask the user for radius
r = input('Enter Radius')
vol = (4/3)*pi*r^3;
disp('Volume is:')
disp(vol)
```

THE WHILE STATEMENT

Anyone who has taken computer science courses has come across the vaunted *while* statement. In MATLAB, we enter while statements by typing:

```
while condition
    statements
end
```

Let's say that we wanted to compute the value of the sum for some given value of n that the user is going to input.

We can do this with a while loop, after getting the cutoff point from the user. First we ask the user how many terms to sum:

```
n = input('Enter number of terms in sum: ')
```

Then we initialize some variables:

```
i = 1;
sum = 0;
```

Now here is the while loop:

```
while i <= n;
sum = sum + 1/I;
i = i + 1;
end
```

We can report the answer to the user this way:

```
disp('Total:')
Total:
disp(sum)
```

One item of business to take note of if you're not a programmer—be sure to increment your counter inside the while loop. This was done with the $i = i + 1$ statement. Otherwise MATLAB will get stuck in an infinite loop.

If we run the code for $n = 5$, then MATLAB tells us the sum is 2.2833.

SWITCH STATEMENTS

Another way to control which statements get executed based on conditions at the time the program is running is to resort to the *switch* statement. The syntax is:

```
switch expression

   case 1
       do these statements
   case 2
       do these statements
   case n
       do these statements
end
```

The case examined can be any type of variable, such as a string or a real number. If multiple cases lead to the same statements being executed, they can be combined into a single case statement by enclosing them on a single comma-delimited line enclosed by parentheses.

Let's cook up a silly example to see how this could work. We work at a government bureaucracy where pay grades are:

```
1, 2, 3, 4
```

With corresponding salaries:

```
$40,000, $65,000, $65,000, and $85,000
```

We can assign salary based on pay grade with a switch statement:

```
switch grade
  case 1
    pay = 40000
  case ( 2,3 )
    pay = 65000
  case 4
    pay = 85000
end
```

Quiz

1. The weights in pounds of a set of male students are described as follows:

Weight	Frequency
130	2
138	1
145	3
150	6
152	1
155	3
160	1
164	1
165	3
167	4
170	3
172	2
175	1

   Use MATLAB to find the average weight, median weight, and standard deviation.

2. Use MATLAB to write a routine to calculate the probability of drawing a given result from a set of frequency weighted data. Using the data in problem 1, what is the probability that a student weighs 150 pounds?
3. Generate a scaled frequency bar chart of the data. From your chart, read off the scaled frequency for 130 pounds.
4. Write a MATLAB routine to ask the user for the radius and height of a cylinder and return the volume.
5. Write a while loop that computes the sum $1 + x + x^2 + \ldots + x^n$.
6. Write a For Loop that computes the sum $1+\frac{1}{x^2}+\frac{1}{x^3}+\cdots+\frac{1}{x^r}$.

CHAPTER 5

Solving Algebraic Equations and Other Symbolic Tools

In this chapter we will begin to look at using MATLAB to solve equations. We start with simple algebraic equations considering solutions for single variables and solving systems of equations. Then we will look at working with transcendental, trig, and hyperbolic functions. Finally we will see how MATLAB handles complex numbers. In the next chapter we will take a look at using MATLAB to solve differential equations.

Solving Basic Algebraic Equations

To solve an algebraic equation in MATLAB we can call upon the *solve* command. At its most basic all we have to do is type in the equation we want to solve enclosed in quotes and hit return. Let's start by looking at a trivial example. Suppose that we wanted to use MATLAB to find the value of x that solves:

$$x + 3 = 0$$

Now the suspense is building—but many clever readers will deduce the answer is $x = -3$ and wonder why we're bothering with this. Well the reason is that it will make seeing how to use MATLAB for symbolic computing a snap. We can find the solution in one step. All we do is create a variable and assign it the value returned by solve in the following way:

```
>> x = solve('x + 3 = 0')

x =

-3
```

Now it isn't necessary to include the right-hand side of the equation. As you can see from the following example, MATLAB *assumes* that when you pass $x + 8$ to solve that you mean $x + 8 = 0$. To verify this, we run this command line:

```
>> x = solve('x+8')

x =

-8
```

So enter the equations whichever way you want. I prefer to be as clear as possible with my intentions, so would rather use $x + 8 = 0$ as the argument.

It is possible to include multiple symbols in the equation you pass to solve. For instance, we might want to have a *constant* included in an equation like this:

$$ax + 5 = 0$$

If we enter the equation in MATLAB, it seems to just assume that we want to solve for x:

```
>> solve('a*x+5')

ans =

-5/a
```

However, there is a second way to call solve. We can tell it *what symbol* we want it to solve for. This is done using the following syntax:

```
solve(equation, variable)
```

Like the equation that you pass to solve, the variable must be enclosed in single quotes. Returning to the equation $ax + 5 = 0$, let's tell MATLAB to find a instead. We do this by typing:

```
>> solve('a*x + 5','a')
```

MATLAB responds with the output:

```
ans =

-5/x
```

Solving Quadratic Equations

The solve command can be used to solve higher order equations, to the delight of algebra students everywhere. For those of us who have moved beyond algebra MATLAB offers us a way to check results when quadratic or cubic equations pop up or to save us from the tedium of solving the equations.

The procedure used is basically the same as we've used so far, we just use a caret character (^) to indicate exponentiation. Let's consider the equation:

$$x^2 - 6x - 12 = 0$$

We could solve it by hand. You can complete the square or just apply the quadratic formula. To solve it using MATLAB, we write:

```
>> s = solve('x^2 -6*x -12 = 0')
```

MATLAB responds with the two roots of the equation:

```
s =

 3+21^(1/2)
 3-21^(1/2)
```

Now, how do you work with the results returned by solve? We can extract them and use them just like any other MATLAB variable. In the case of two roots like this one, the roots are stored as $s(1)$ and $s(2)$. So we can use one of the roots to define a new quantity:

```
>> y = 3 + s(1)

y =

6+21^(1/2)
```

Here is another example where we refer to both roots returned by solve:

```
>> s(1) + s(2)

ans =

6
```

When using solve to return a single variable, the array syntax is not necessary. For example, we use x to store the solution of the following equation:

```
>> x = solve('3*u + 9 = 8')
```

MATLAB dutifully tells us that:

```
x =

-1/3
```

Now we can use the result in another equation:

```
>> z = x + 1

z =

2/3
```

It is possible to assign an equation to a variable and then pass the variable to solve. For instance, let's create an equation (generated at random on the spot) and assign it to a variable with the meaningless name d:

```
>> d = 'x^2 + 9*x -7 = 0';
```

Now we call solve this way:

```
>> solve(d)
```

MATLAB correctly tells us the two roots of the equation:

```
ans =

 -9/2+1/2*109^(1/2)

 -9/2-1/2*109^(1/2)
```

Plotting Symbolic Equations

Let's take a little detour from solving equations, to see how we can plot symbolically entered material. OK while I argued that writing 'x^2 – 6*x – 12 = 0' was better than 'x^2 – 6*x – 12' it turns out there is a good reason why you might choose the latter method. The reason is that MATLAB allows us to generate plots of symbolic equations we've entered. This can be done using the *ezplot* command. Let's do that for this example.

First let's create the string to represent the equation:

```
>> d = 'x^2 -6*x - 12';
```

Now we call ezplot:

```
>> ezplot(d)
```

MATLAB responds with the graph shown in Figure 5-1.

A couple of things to notice are:

- ezplot has conveniently generated a title for the plot shown at the top, without us having to do any work.
- It has labeled the x axis for us.

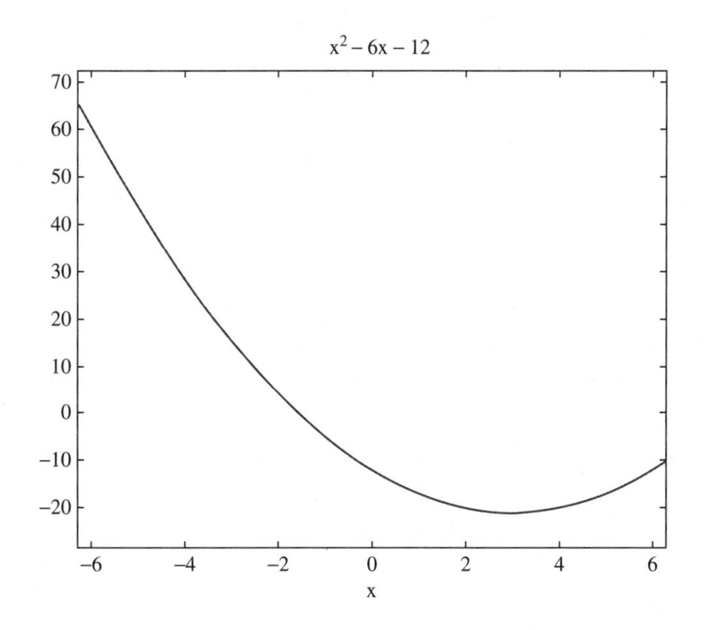

Figure 5-1 A plot of a symbolic function generated using ezplot

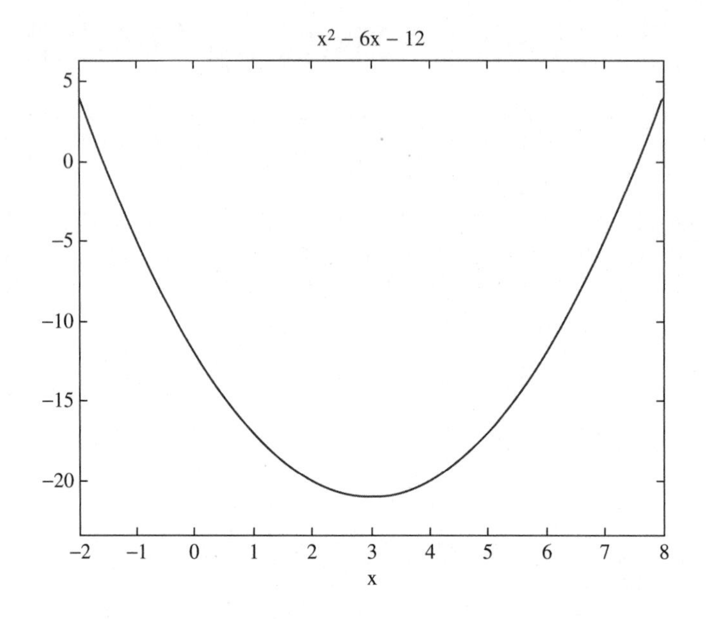

Figure 5-2 Using ezplot while specifying the domain

The function also picked what values to use for the domain and range in the plot. Of course we may not like what it picked. We can specify what we want by specifying the domain with the following syntax:

```
ezplot(f, [x1, x2])
```

This plots f for $x_1 < x < x_2$. Returning to the previous example, let's say we wanted to plot it for $-2 < x < 8$. We can do that using the following command:

```
>> d = 'x^2 -6*x - 12';
>> ezplot(d,[-2,8])
```

The plot generated this time is shown in Figure 5-2.

Before we go any further, let's get back to the " = 0" issue. Suppose we tried to plot:

```
>> ezplot('x+3=0')
```

MATLAB doesn't like this at all. It spits out a series of meaningless error messages:

```
??? Error using ==> inlineeval
Error in inline expression ==> x+3=0
```

```
??? Error: The expression to the left of the equals sign is
not a valid target for an assignment.

Error in ==> inline.feval at 34
        INLINE_OUT_ = inlineeval(INLINE_INPUTS_, INLINE_OBJ_
.inputExpr, INLINE_OBJ_.expr);

Error in ==> specgraph\private\ezplotfeval at 54
    z = feval(f,x(1));

Error in ==> ezplot>ezplot1 at 448
[y,f,loopflag] = ezplotfeval(f,x);

Error in ==> ezplot at 148
    [hp,cax] = ezplot1(cax,f{1},vars,labels,args{:});
```

Now, if we instead type:

```
>> ezplot('x+3')
```

It happily generates the straight line shown in Figure 5-3.

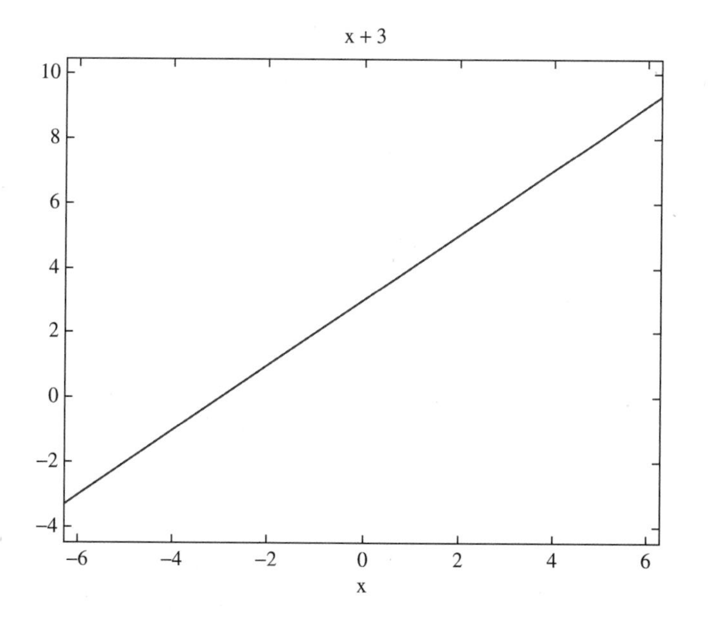

Figure 5-3 Using ezplot('x + 3') works, but *ezplot*('x + 3 = 0') will generate an error

Now we just mentioned a while ago that we could tell ezplot how to specify the domain to include in the plot. Naturally it also allows us to specify the range. Just for giggles, let's say we wanted to plot:

$$x + 3 = 0$$
$$-4 < x < 4, -2 < y < 2$$

We can do this by typing:

```
>> ezplot('x+3',[-4,4,-2,2])
```

The plot generated is shown in Figure 5-4. So, to specify you want the plot to be over $x_1 < x < x_2$ and $y_1 < y < y_2$ include $[x_1, x_2, y_1, y_1]$ in your call to ezplot.

EXAMPLE 5-1
Find the roots of $x^2 + x - \sqrt{2} = 0$ and plot the function. Determine the numerical value of the roots.

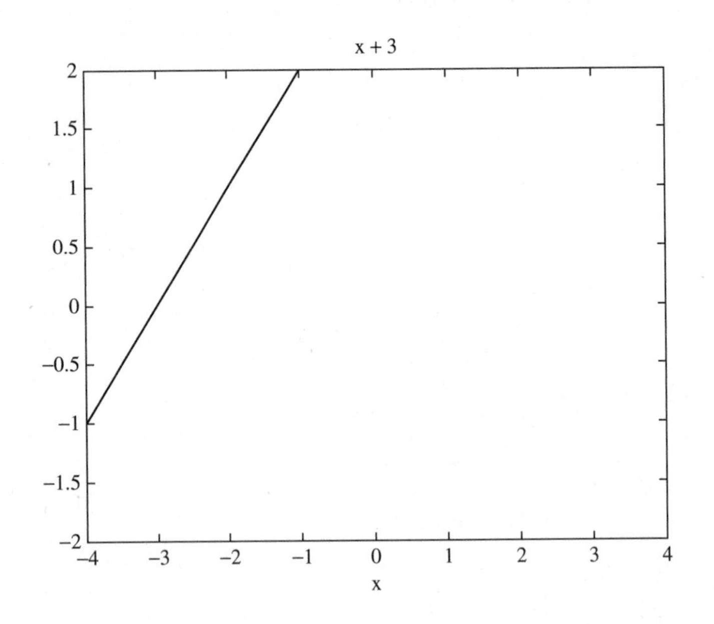

Figure 5-4 Using ezplot with the call ezplot('x + 3', [–4, 4, –2, 2])

SOLUTION 5-1

First let's create a string to represent the equation. First as an aside, note that you *can* include predefined MATLAB expressions in your equation. So it would be perfectly OK to enter the equation as:

```
>> eq = 'x^2 + x - sqrt(2)';
```

Or if you like, you could write:

```
>> eq = 'x^2 + x - 2^(1/2)';
```

Next we call solve to find the roots:

```
>> s = solve(eq)

s =
 -1/2+1/2*(1+4*2^(1/2))^(1/2)
 -1/2-1/2*(1+4*2^(1/2))^(1/2)
```

To determine the numerical value of the roots, we need to extract them from the array and convert them into type double. This is done by simply passing them to the double(.) command. For example, we get the first root out by writing:

```
>> x = double(s(1))

x =

    0.7900
```

And the second root:

```
>> y = double(s(2))

y =

   -1.7900
```

To plot the function, we use a call to ezplot:

```
>> ezplot(eq)
```

The result is shown in Figure 5-5.

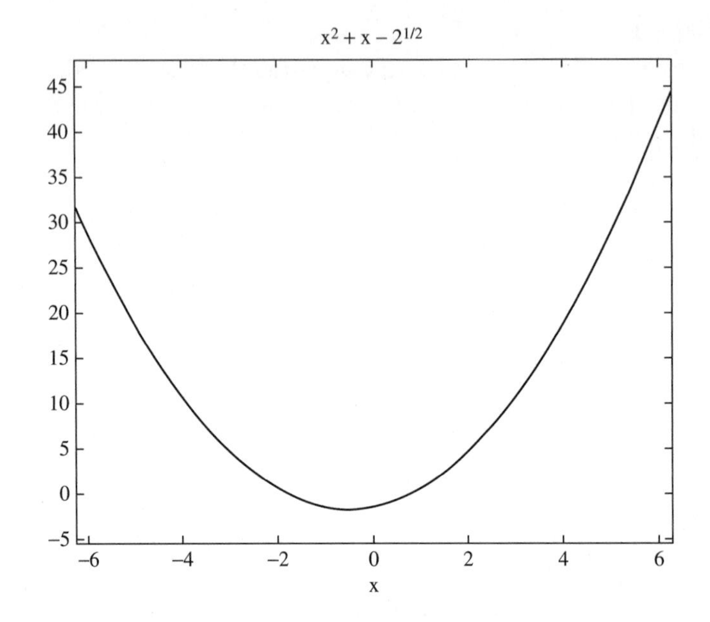

Figure 5-5 A plot of the quadratic equation solved in Example 5-1

Solving Higher Order Equations

Of course we can use MATLAB to solve higher order equations. Let's try a cubic. Suppose we are told that:

$$(x + 1)^2 (x - 2) = 0$$

Solving an equation like this is no different than what we've done so far. We find that the roots are:

```
>> s = solve(eq)

s =

  2
 -1
 -1
```

EXAMPLE 5-2

Find the roots of the fourth order equation

$$x^4 - 5x^3 + 4x^2 - 5x + 6 = 0$$

and plot the function for $-10 < x < 10$.

SOLUTION 5-2

First we define the function by creating a character string to represent it:

```
>> eq1 = 'x^4-5*x^3+4*x^2-5*x+6';
```

Then we call solve to find the roots:

```
>> s = solve(eq1);
```

Now let's define some variables to extract the roots from *s*. If you list them symbolically, you will get a big mess. We show part of the first root here:

```
>> a = s(1)

a =

5/4+1/12*3^(1/2)*((43*(8900+12*549093^(1/2))^(1/3)+2*(8900+
12*549093^(1/2))^(2/3)+104)....
```

Try it and you will see this term goes on a long way. So let's use double to get a numerical result:

```
>> a = double(s(1))

a =

    4.2588
```

Now that is much nicer. We get the rest of the roots. Since it's a fourth order equation, well there are four roots:

```
>> b = double(s(2))

b =

    1.1164
```

```
>> c = double(s(3))

c =

  -0.1876 + 1.1076i

>> d = double(s(4))

d =

  -0.1876 - 1.1076i
```

Notice two of the roots are complex numbers. Now let's plot the function over the domain indicated:

```
>> ezplot(eq1,[-10 10])
```

The result is shown in Figure 5-6.

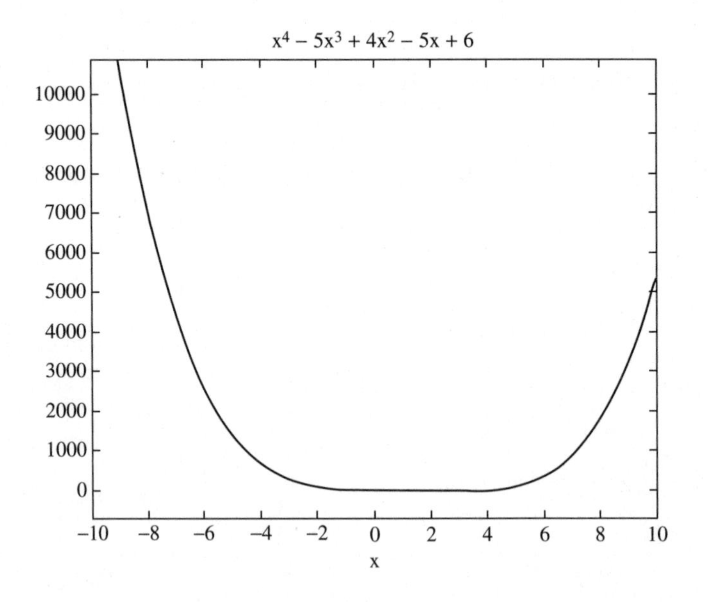

Figure 5-6 Plot of the function $x^4 - 5x^3 + 4x^2 - 5x + 6$

EXAMPLE 5-3

Find the roots of $x^3 + 3x^2 - 2x - 6$ and plot the function for $-8 < x < 8$, $-8 < y < 8$. Generate the plot with a grid.

SOLUTION 5-3

Again we follow the usual steps. First define the function as a character string:

```
>> f = 'x^3+3*x^2-2*x-6';
```

Calling solve, we find that the roots are -3 and $\pm\sqrt{2}$:

```
>> s = solve(f)

s =

       -3
   2^(1/2)
  -2^(1/2)
```

Now we call ezplot to generate a plot of the function, with the specification that $-8 < x < 8$, $-8 < y < 8$, and we add the command grid on:

```
>> ezplot(f,[-8,8,-8,8]), grid on
```

The result is shown in Figure 5-7.

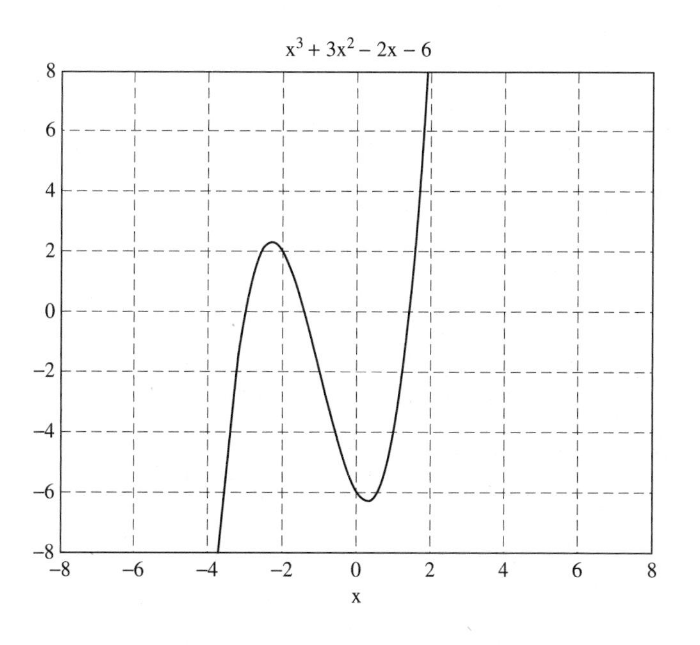

Figure 5-7 A plot of $x^3 + 3x^2 - 2x - 6$

Systems of Equations

While appearing useful already, it turns out the solve command is more versatile than we have seen so far. It turns out that solve can be used to generate solutions of systems of equations. To show how this is done and how to get the solutions out, it's best to proceed with another simple example. Suppose that you were presented with the following system of equations:

$$5x + 4y = 3$$
$$x - 6y = 2$$

To use solve to find x and y, we call it by passing two arguments—each one a string representing one of the equations. In this case we type:

```
>> s = solve('5*x + 4*y = 3','x - 6*y = 2');
```

Notice that each equation is enclosed in single quote marks and that we use a comma to delimit the equations. We can get the values of x and y by using a "dot" notation as follows. First let's get x:

```
>> x = s.x

x =

13/17
```

MATLAB has generated a nice fraction for us that we can write in on our homework solutions—the professor will never suspect a thing. Next we get y:

```
>> y = s.y

y =

-7/34
```

This method can be used with larger linear systems. Consider:

$$w + x + 4y + 3z = 5$$
$$2w + 3x + y - 2z = 1$$
$$w + 2x - 5y + 4z = 3$$
$$w - 3z = 9$$

We can enter the system as a set of character strings:

```
>> eq1 = 'w + x + 4*y + 3*z = 5';
>> eq2 = '2*w +3*x+y-2*z = 1';
>> eq3 = 'w + 2*x-5*y+4*z = 3';
>> eq4 = 'w - 3*z = 9';
```

Now we call solve, passing the equations as a comma-delimited list:

```
>> s = solve(eq1,eq2,eq3,eq4);
```

We can then extract the value of each variable which solves the system of equations:

```
>> w = s.w

w =

1404/127

>> x = s.x

x =

-818/127

>> y = s.y

y =

-53/127

>> z = s.z

z =

87/127
```

Expanding and Collecting Equations

In elementary school we learned how to expand equations. For instance:

$$(x + 2)\,(x - 3) = x^2 - x - 6$$

We can use MATLAB to accomplish this sort of task by calling the *expand* command. Using expand is relatively easy. For example:

```
>> expand((x-1)*(x+4))
```

When this command is executed, we get:

```
ans = x^2 +3*x - 4
```

The expand function can be applied in other ways. For example, we can apply it to trig functions, generating some famous trig identities:

```
>> expand(cos(x+y))
```

This gives us:

```
ans =

cos(x)*cos(y)-sin(x)*sin(y)
```

Or:

```
>> expand(sin(x-y))

ans =

sin(x)*cos(y)-cos(x)*sin(y)
```

To work with many symbolic functions, you must tell MATLAB that your variable is symbolic. For example, if we type:

```
>> expand((y-2)*(y+8))
```

MATLAB returns:

```
??? Undefined function or variable 'y
```

To get around this, first enter:

```
>> syms y
```

Then we get:

```
>> expand((y-2)*(y+8))

ans =

y^2+6*y-16
```

MATLAB also lets us go the other way, collecting and simplifying equations. First let's see how to use the *collect* command. One way you can use it is for distribution of multiplication. Consider:

$$x(x^2 - 2) = x^3 - 2x$$

To do this in MATLAB, we just write:

```
>> collect(x*(x^2-2))

ans =

x^3-2*x
```

Another example (with a new symbolic variable t):

```
>> syms t
>> collect((t+3)*sin(t))

ans =

sin(t)*t+3*sin(t)
```

Another algebraic task we can do symbolically is factoring. To show that:

$x^2 - y^2 = (x + y)(x - y)$

Using MATLAB, we type:

```
>> factor(x^2-y^2)

ans =

(x-y)*(x+y)
```

We can factor multiple equations with a single command:

```
>> factor([x^2-y^2, x^3+y^3])
```

Finally, we can use the *simplify* command. This command can be used to divide polynomials. For instance, we can show that $(x^2 + 9)(x^2 - 9) = x^4 - 81$ by writing:

```
>> simplify((x^4-81)/(x^2-9))

ans =

x^2+9
```

Here is another example. Consider that:

$$e^{2\log 3x} = e^{\log 9x^2} = 9x^2$$

We could have found this out with the MATLAB command:

```
>> simplify(exp(2*log(3*x)))

ans =

9*x^2
```

The simplify command is useful for obtaining trig identities. Examples:

```
>> simplify(cos(x)^2-sin(x)^2)

ans =

2*cos(x)^2-1

>> simplify(cos(x)^2+sin(x)^2)

ans =

1
```

Solving with Exponential and Log Functions

So far for the most part we have only looked at polynomials. The symbolic solver can also be used with exponential and logarithmic functions. First let's consider and equation with logarithms.

EXAMPLE 5-4
Find a value of x that satisfies:

$$\log_{10}(x) - \log_{10}(x-3) = 1$$

SOLUTION 5-4
Base ten logarithms can be calculated by or represented by the *log*10 function in MATLAB. So we can enter the equation thus:

```
>> eq = 'log10(x)-log10(x-3) = 1';
```

Then we call solve:

```
>> s = solve(eq);
```

In this case, MATLAB only returns one solution:

```
>> s(1)

ans =

10/3
```

Now let's consider some equations involving variables as exponents. Suppose we are asked to solve the system:

$$y = 3^{2x}$$
$$y = 5^x + 1$$

We can call solve to find the solution:

```
>> s = solve('y = 3^2*x','y = 5^x+1')
```

MATLAB returns:

```
s =

    x: [2x1 sym]
    y: [2x1 sym]
```

So it appears there are two values of each variable that solve the equation. We can get the values of x out by typing:

```
>> s.x(1)

ans =

1/9*exp(-lambertw(-1/9*log(5)*5^(1/9))+1/9*log(5))+1/9

>> s.x(2)

ans =

1/9*exp(-lambertw(-1,-1/9*log(5)*5^(1/9))+1/9*log(5))+1/9
```

These might not be too useful to the average man or woman. So let's convert them to numerical values:

```
>> a = double(s.x(1))

a =

    0.2876

>> b = double(s.x(2))

b =

    1.6214
```

We can also extract the y values:

```
>> c = double(s.y(1))

c =

    2.5887

>> d = double(s.y(2))

d =

   14.5924
```

Do the results make sense? We check. The idea is to see if s.x(1) satisfies the first equation giving s.y(1) and ditto for the second array elements. We find:

```
>> 3^2*a

ans =

    2.5887
```

This agrees with s.y(1), so far so good. Now:

```
>> 5^b+1

ans =

   14.5924
```

Hence, s.y(2) is consistent with s.x(2) and we have the solution.

We can also enter and solve equations involving the exponential function. For example:

```
>> eq = 'exp(x)+ x';
>> s = solve(eq)

s =

-lambertw(1)
```

If you're not familiar with Lambert's *w* function, you can evaluate numerically:

```
>> double(s)

ans =

   -0.5671
```

We can plot the function using ezplot:

```
>> double(s)
```

The result is shown in Figure 5-8.

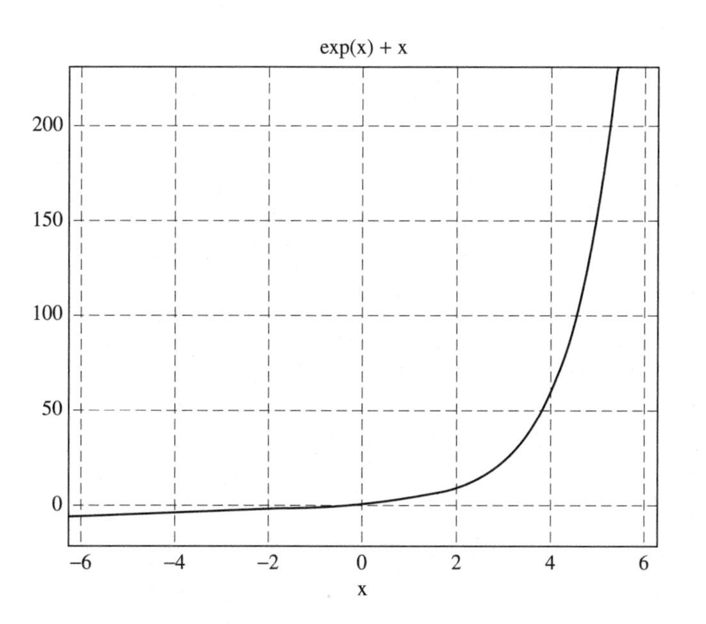

Figure 5-8 A plot of $e^x + x$

Series Representations of Functions

We close out the chapter with a look at how MATLAB can be used to obtain the series representation of a function, given symbolically. The *taylor* function returns the Taylor series expansion of a function. The simplest way is to call taylor directly. For example, we can get the Taylor series expansion of sin(x):

```
>> syms x
>> s = taylor(sin(x))

s =

x-1/6*x^3+1/120*x^5
```

MATLAB has returned the first three terms of the expansion. In fact what MATLAB returns is:

$$\sum_{n=0}^{5} \frac{f^{(n)}(0)}{n!} x^n$$

In the case of sin, these are the only nonzero terms. Let's plot it:

```
>> ezplot(s)
```

The plot is shown in Figure 5-9.

This might be an accurate representation of the function for small values, but it clearly doesn't look much like sin(x) over a very large domain. To get MATLAB to return more terms, say we want to approximate the function by m terms:

$$\sum_{n=0}^{m} \frac{f^{(n)}(0)}{n!} x^n$$

We can make the call by writing taylor(f, m). For the present example, let's try 20 terms:

```
> s = taylor(sin(x),20)

s =

x-1/6*x^3+1/120*x^5-1/5040*x^7+1/362880*x^9-1/
39916800*x^11+1/6227020800*x^13-1/1307674368000*x^15+1/
355687428096000*x^17-1/121645100408832000*x^19
```

The representation this time is far more accurate, as the plot in Figure 5-10 shows.

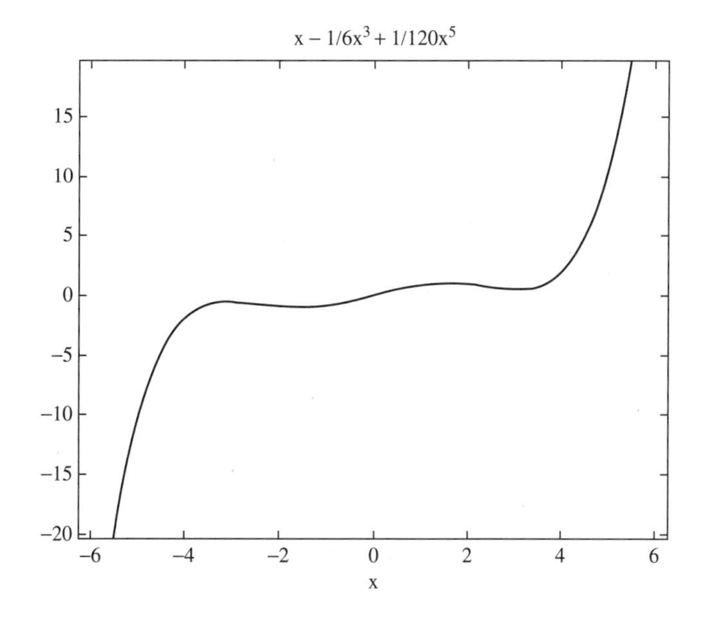

Figure 5-9 Plot of the Taylor series expansion of sin(x)

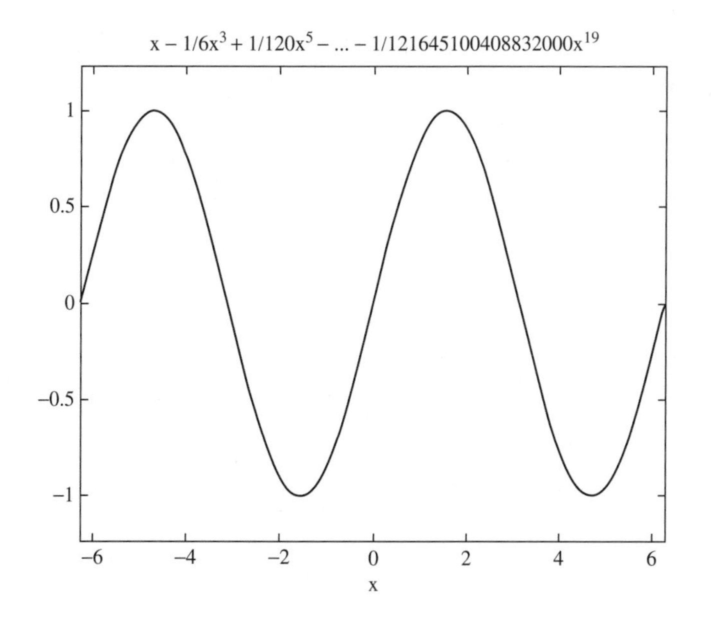

Figure 5-10 When we generate the Taylor series expansion of the sin function with 20 terms, we get a more accurate representation

Quiz

1. Use MATLAB to enter $7\sqrt{2}-5\sqrt{60}+5\sqrt{8}$ as a string, then find the numerical value.
2. Use MATLAB to solve $3x^2 + 2x = 7$.
3. Find x such that $x^2+\sqrt{5}x-\pi=0$.
4. Find the solution of $\sqrt{2x-4}=1$ and symbolically plot the function for $2 < x < 4, 0 < y < 1$.
5. Use solve to symbolically find the roots of $2t^3 - t^2 + 4t - 6 = 0$, then convert the answer into numerical values.
6. Find a solution of the system:

$$\begin{aligned} x - 3y - 2z &= 6 \\ 2x - 4y - 3z &= 8 \\ -3x + 6y + 8z &= -5 \end{aligned}$$

7. Does the equation $e^x - x^2 = 0$ have a real root?
8. Use MATLAB to find $\tan^2 x - \sec^2 x$.
9. Find the tenth order Taylor expansion of $\tan(x)$.
10. Find the tenth order Taylor series expansion of $\frac{4}{5-\cos x}$. Does your result agree with the plot shown in Figure 5-11?

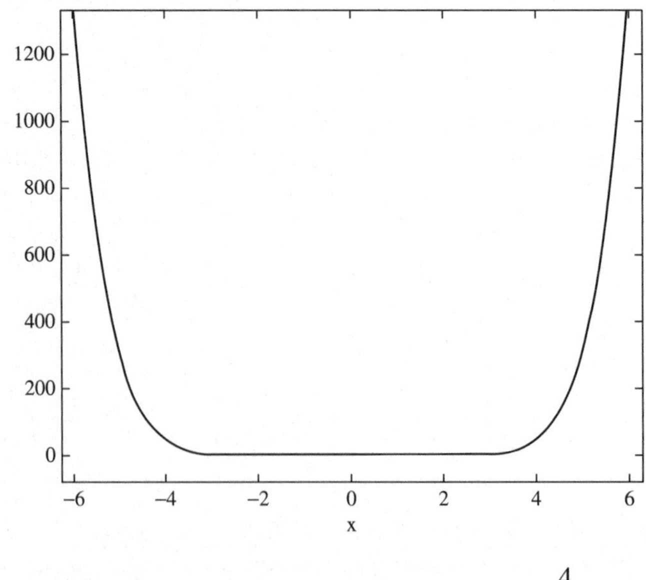

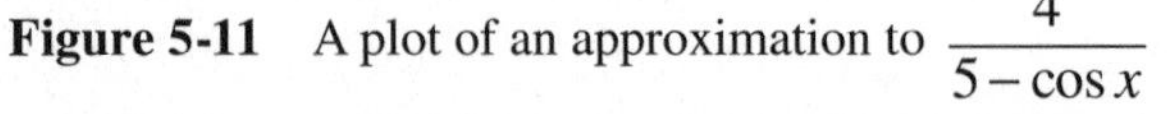
Figure 5-11 A plot of an approximation to $\frac{4}{5-\cos x}$

CHAPTER 6

Basic Symbolic Calculus and Differential Equations

In this chapter we will learn how to use MATLAB to do symbolic calculus. Specifically, we will start by examining limits and derivatives, and then see how to solve differential equations. We will cover integration in the next chapter.

Calculating Limits

MATLAB can be used to calculate limits by making a call to the *limit* command. The most basic way to use this command is to type in the expression you want to use. MATLAB will then find the limit of the expression as the independent variable goes to zero. For example, if you enter a function $f(x)$, MATLAB will then find $\lim_{x \to 0} f(x)$. Here is an example:

```
>> syms x
>> limit((x^3 + 1)/(x^4 + 2))

ans =

1/2
```

Remember, the limit command falls in the realm of symbolic computing, so be sure to use the *syms* command to tell MATLAB which symbolic variables you are using.

To compute $\lim_{x \to a} f(x)$, the limit command is called using the syntax *limit(f, a)*. For example:

```
>> limit(x + 5,3)

ans =

8
```

EXAMPLE 6-1

Let $f(x) = \frac{2x+1}{x-2}$ and $g(x) = x^2 + 1$. Compute the limit as $x \to 3$ of both functions and verify the basic properties of limits using these two functions and MATLAB.

SOLUTION 6-1

First we tell MATLAB what symbolic variables we will use and define the functions:

```
>> syms x
>> f = (2*x + 1)/(x-2);
>> g = x^2 + 1;
```

Now let's find the limit of each function, and store the result in a variable we can use later:

```
>> F1 = limit(f,3)

F1 =

7


>> F2 = limit(g,3)

F2 =

10
```

The first property of limits we wish to verify is:

$$\lim_{x\to a}(f(x)+g(x)) = \lim_{x\to a} f(x) + \lim_{x\to a} g(x)$$

From our calculations so far we see that:

$$\lim_{x\to 3} f(x) + \lim_{x\to 3} g(x) = 7 + 10 = 17$$

Now let's verify the relation by calculating the left-hand side:

```
>> limit(f+g,3)

ans =

17
```

Next we can verify:

$$\lim_{x\to a} k\, f(x) = k \lim_{x\to a} f(x)$$

for any constant k. Let's let $k = 3$ for which we should find:

$$\lim_{x\to a} k\, f(x) = k \lim_{x\to a} f(x) = (3)(7) = 21:$$

```
>> k=3;
>> limit(k*f,3)

ans =

21
```

Now let's check the fact that the limit of the product of two functions is the product of their limits, that is:

$$\lim_{x\to a} f(x)g(x) = \lim_{x\to a} f(x) \lim_{x\to a} g(x)$$

The product of the limits is:

```
>> F1*F2

ans =

70
```

And we find the limit of the product to be:

```
>> limit(f*g,3)

ans =

70
```

Finally, let's verify that:

$$\lim_{x\to a} f(x)^{g(x)} = \left(\lim_{x\to a} f(x)\right)^{\lim_{x\to a} g(x)}$$

We can create $f(x)^{g(x)}$ in MATLAB:

```
>> h = f^g

h =

((2*x+1)/(x-2))^(x^2+1)
```

Computing the limit:

```
>> limit(h,3)

ans =

282475249
```

Checking the right side of the relation, we find that they are equal:

```
>> A = F1^F2

A =

282475249
```

As an aside, we can check if two quantities in MATLAB are equal by calling the *isequal* command. If two quantities are not equal, isequal returns 0. Recall that earlier we defined a constant $k = 3$. Here is what MATLAB returns if we compare it to A = F1^F2:

```
>> isequal(A,k)

ans =

     0
```

On the other hand:

```
>> isequal(A,limit(h,3))

ans =

     1
```

Computing $\lim_{x \to \infty} f(x)$.

We can calculate limits of the form $\lim_{x \to \infty} f(x)$ by using the syntax:

```
limit(f,inf)
```

Let's use MATLAB to show that $\lim_{x \to \infty}\left(\sqrt{x^2 + x} - x\right) = \frac{1}{2}$:

```
>> limit(sqrt(x^2+x)-x,inf)

ans =

1/2
```

We can also calculate $\lim_{x \to -\infty} f(x)$. For example:

```
>> limit((5*x^3 + 2*x)/(x^10 + x + 7),-inf)

ans =

0
```

MATLAB will also tell us if the result of a limit is ∞. For example, we verify that $\lim_{x \to 0} \frac{1}{|x|} = \infty$:

```
>> limit(1/abs(x))

ans =

Inf
```

LEFT- AND RIGHT-SIDED LIMITS

When a function has a discontinuity, the limit does not exist at that point. To handle limits in the case of a discontinuity at $x = a$, we define the notion of *left-handed* and *right-handed* limits. A left-handed limit is defined as the limit as $x \to a$ from the left, that is x approaches a for values of $x < a$. In calculus we write:

$$\lim_{x \to a^-} f(x)$$

For a right-handed limit, where $x \to a$ from the right, we consider the case when x approaches a for values of $x > a$. The notation used for right-handed limits is:

$$\lim_{x \to a^+} f(x)$$

If these limits are equal, then $\lim_{x \to a} f(x)$ exists. In MATLAB, we can compute left- and right-handed limits by passing the character strings 'left' and 'right' to the limit command as the last argument. We must also tell MATLAB the variable we are using to compute the limit in this case. Let's illustrate with an example.

EXAMPLE 6-2

Show that $\lim_{x \to 3} \frac{x-3}{|x-3|}$ does not exist.

SOLUTION 6-2

First let's define the function in MATLAB:

```
>> f = (x - 3)/abs(x-3);
```

If we plot the function, the discontinuity at $x = 3$ is apparent, as shown in Figure 6-1. Note that we have to give MATLAB the domain over which we want to plot in order to get it to show us the discontinuity:

```
>> ezplot(f,[-1,5])
```

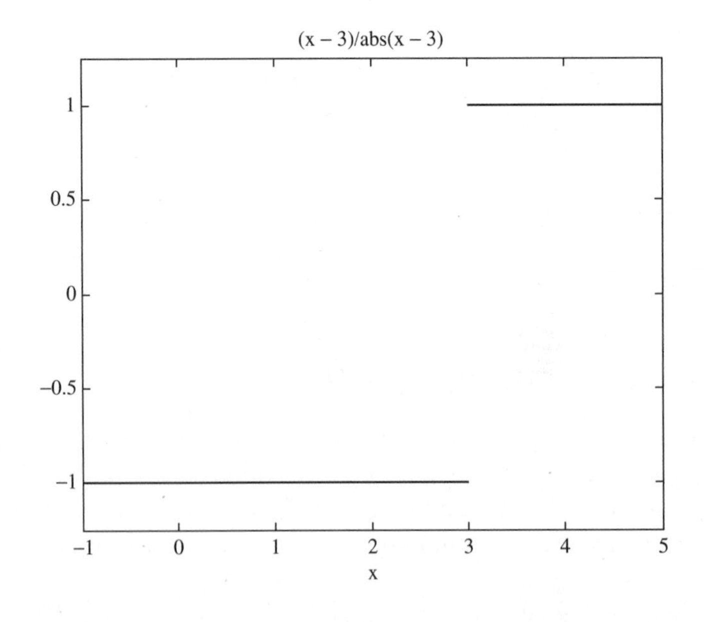

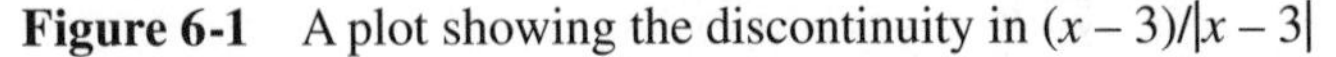
Figure 6-1 A plot showing the discontinuity in $(x - 3)/|x - 3|$

Now let's compute the left-handed limit. To do this, we must pass the function, the variable we are using to take the limit, and the string 'left' as a comma-delimited list:

```
>> a = limit(f,x,3,'left')

a =

-1
```

Now let's take the right-handed limit:

```
>> b = limit(f,x,3,'right')

b =

1
```

Since these two terms are not equal, we have shown that $\lim_{x\to 3}\frac{x-3}{|x-3|}$ does not exist.

FINDING ASYMPTOTES

The limit command can be used to find the asymptotes of a function. Let's see how we can find the asymptotes and generate a plot showing the function together with its asymptotes.

For our example, let's consider the function:

$$y = \frac{1}{x(x-1)}$$

We choose this example because it's pretty clear where the asymptotes are—the function is going to blow up when $x = a$ and $x = 1$. Let's generate a quick plot of the function:

```
>> f = 1/(x*(x-1));
>> ezplot(f)
```

The result is shown in Figure 6-2, and you can see the function veering off to infinity at the appropriate points. Now let's develop a formal process to find out what those points are (we are pretending we don't know) and then show them on the graph.

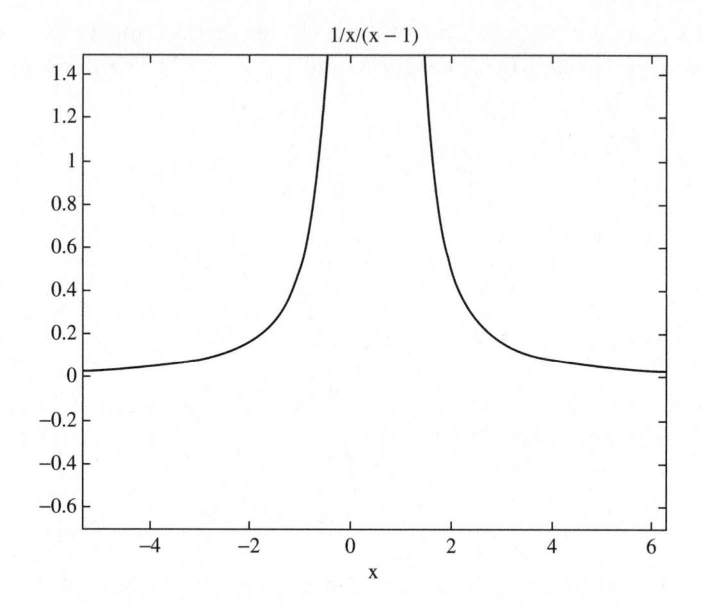

Figure 6-2 A plot of a function that blows up at two points

The first step is to find the points where the function blows up. This can be done by finding the roots of the denominator. Let's create a function to represent the denominator and then find the roots. Thinking back to the last chapter, we can use the solve command:

```
>> g = x*(x-1);
>> s = solve(g)

s =

 0
 1
```

With the roots in hand, we know where the asymptotes are. We will draw them as dashed lines. First we plot the function again, and then use the hold on command to tell MATLAB we are going to add more material to the plot:

```
>> ezplot(1/g)
>> hold on
```

Since we stored the roots in a variable called *s*, remember that we access each of them by writing *s*(1) and *s*(2). We can plot the first asymptote using the following command:

```
>> plot(double(s(1))*[1 1], [-1 2],'--')
```

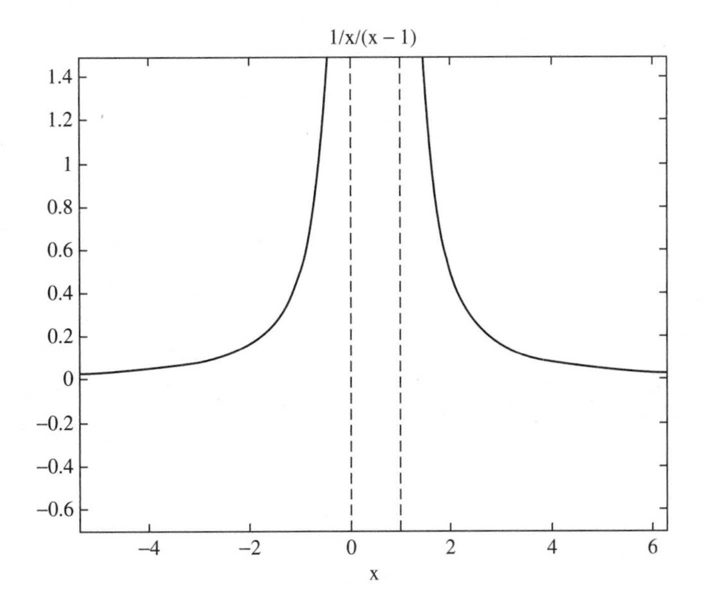

Figure 6-3 A plot of the function shown in Figure 6-2 together with its asymptotes

The range [–1 2] gives the range of the *y* axis over which we want to draw the line. Try playing around with different values to see how it works. Now let's draw a dashed line for the second asymptote:

```
>> plot(double(s(2))*[1 1], [-1 2],'--')
>> hold off
```

We call the hold off command to free MATLAB from having to continue plotting to the same graph. The result is shown in Figure 6-3.

Computing Derivatives

We can compute symbolic derivatives using MATLAB with a call to the *diff* command. Simply pass the function you want to differentiate to diff as this example shows:

```
>> syms x t
>> f = x^2;
>> g = sin(10*t);
>> diff(f)
```

```
ans =

2*x


>> diff(g)

ans =

10*cos(10*t)
```

To take higher derivatives of a function *f*, we use the syntax diff(*f*,*n*). Let's find the second derivative of *t* exp (–3*t*):

```
>> f = t*exp(-3*t);
>> diff(f,2)

ans =

-6*exp(-3*t)+9*t*exp(-3*t)
```

As you can see, diff returns the result of calculating the derivative—so we can assign the result to another variable that can be used later.

EXAMPLE 6-3

Show that $f(x) = x^2$ satisfies $-\frac{df}{dx} + 2x = 0$.

SOLUTION 6-3

We start by making some definitions:

```
>> syms x
>> f = x^2; g = 2*x;
```

Now let's compute the required derivative:

```
>> h = diff(f);
```

Finally, verify that the relation is satisfied:

```
>> -h+g

ans =

0
```

EXAMPLE 6-4

Does $y(t) = 3 \sin t + 7 \cos 5t$ solve $y'' + y = -5 \cos 2t$?

SOLUTION 6-4

Define our function:

```
>> y = 3*sin(t)+7*cos(5*t);
```

Now let's create a variable to hold the required result:

```
>> f = -5*cos(2*t);
```

To enter the left-hand side of the differential equation, we create another variable:

```
>> a = diff(y,2)+y;
```

We use isequal to check whether the equation is satisfied:

```
>> isequal(a,f)

ans =

     0
```

Since 0 is returned, $y(t) = 3 \sin t + 7 \cos 5t$ does not solve $y'' + y = -5 \cos 2t$.

EXAMPLE 6-5

Find the minima and maxima of the function $f(x) = x^3 - 3x^2 + 3x$ in the interval $[0, 2]$.

SOLUTION 6-5

First let's enter the function and plot it over the given interval:

```
>> syms x
>> f = x^3-3*x^2+3*x;
>> ezplot(f, [0 2])
```

The plot is shown in Figure 6-4. To find local maxima and minima, we will compute the derivative and find the points at which it vanishes.

The derivative is:

```
>> g = diff(f)

g =

3*x^2-6*x+3
```

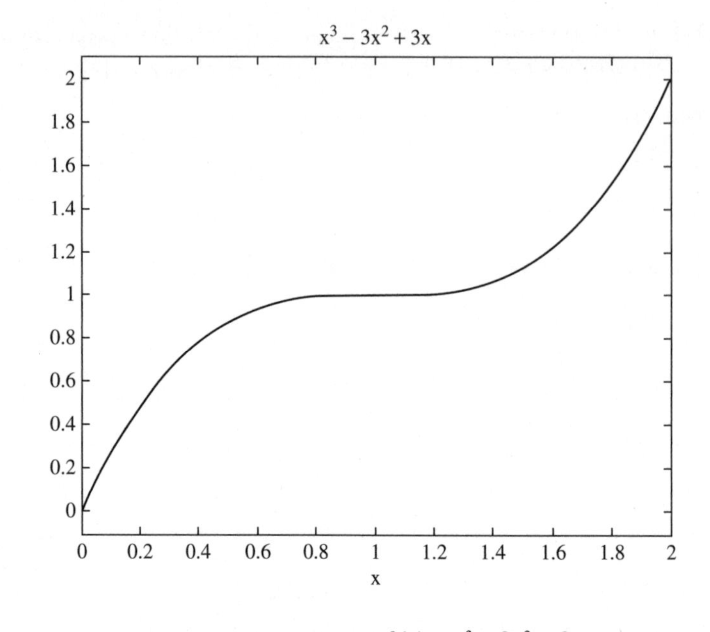

Figure 6-4 A plot of $f(x) = x^3 - 3x^2 + 3x$

As a quick aside, we can use the pretty command to make our expressions look nicer:

```
>> pretty(g)

                                    2
                                 3 x  - 6 x + 3
```

Well slightly nicer, anyway. Returning to the problem, let's set the derivative equal to zero and find the roots:

```
>> s = solve(g)

s =

  1
  1
```

We see that there is only one critical point, since the derivative has a double root. We can see from the plot that the maximum occurs at the endpoint, but let's prove this by evaluating the function at the critical points $x = 0, 1, 2$.

We can substitute a value in a symbolic function by using the *subs* command. With a single variable this is pretty simple. If we want to set $x = c$, we make the call

subs(*f*,*c*). So let's check *f* for $x = 0, 1, 2$. We can check all three on a single line and have MATLAB report the output by passing a comma-delimited list:

```
>> subs(f,0), subs(f,1), subs(f,2)

ans =

     0


ans =

     1


ans =

     2
```

Since $f(2)$ returns the largest value, we conclude that the maximum occurs at $x = 0$. For fun, let's evaluate the derivative at these three points and plot it:

```
>> subs(g,0),subs(g,1),subs(g,2)

ans =

     3


ans =

     0


ans =

     3
```

Where are the critical points of the derivative? We take the second derivative and set equal to zero:

```
>> h = diff(g)

h =
```

```
6*x-6

>> solve(h)

ans =

1
```

The next derivative is:

```
>> y = diff(h)

y =

6
```

Since $g'' > 0$ we can conclude that the critical point $x = 1$ is a local minimum. A plot of $g(x)$ is shown in Figure 6-5.

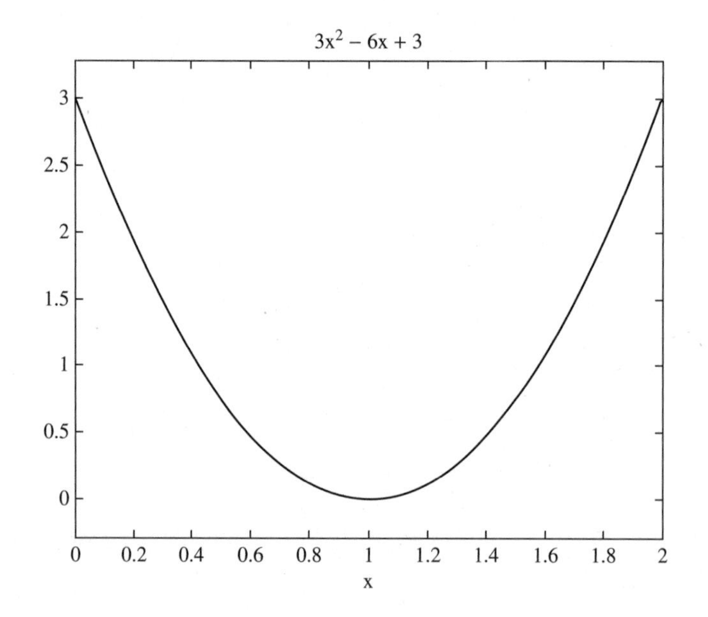

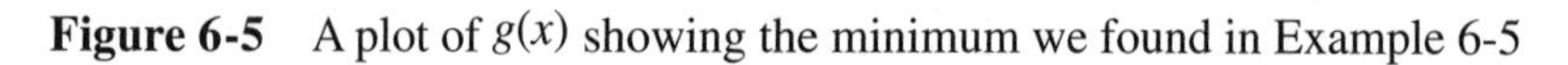

Figure 6-5 A plot of $g(x)$ showing the minimum we found in Example 6-5

EXAMPLE 6-6

Plot the function $f(x) = x^4 - 2x^3$ and show any local minima and maxima.

SOLUTION 6-6

First we define the function and plot it:

```
>> clear
>> syms x
>> f = x^4-2*x^3;
>> ezplot(f,[-2 3])
>> hold on
```

Now compute the first derivative:

```
>> g = diff(f)

g =

4*x^3-6*x^2
```

We find the critical numbers by setting the first derivative equal to zero and solving:

```
>> s = solve(g)

s =

 3/2
   0
   0
```

Next we compute the second derivative:

```
>> h = diff(g)

h =

12*x^2-12*x
```

Evaluating at the first critical number which is $x = 3/2$ we find:

```
>> a = subs(h,s(1))

a =

9
```

Since $f''(3/2) = 9 > 0$, the point $x = 3/2$ is a local minimum. Now let's check the other critical number:

```
>> b = subs(h,s(2))

b =

0
```

In this case, $f''(0) = 0$, which means that we cannot get any information about the point using the second derivative test. It can be shown easily that the point is neither a minimum nor a maximum. Now let's fix up the plot to show the local minimum. We add the point $(c, f(c))$ where $c = s(1)$ to the plot using the following command:

```
>> plot(double(s(1)),double(subs(f,s(1))),'ro
```

This puts a small red circle at the point $(3/2, f(3/2))$ on the plot. We told MATLAB to make the circle red by entering 'ro' instead of 'o', which would have added a black circle. Now let's label the point as a local minimum. This can be done using the text command. When you call text, you pass it the *x-y* coordinates where you want the text to start printing, and then pass the text string you want to appear on the plot:

```
>> text(0.8,3.2,'Local minimum')
>> hold off
```

The result is shown in Figure 6-6.

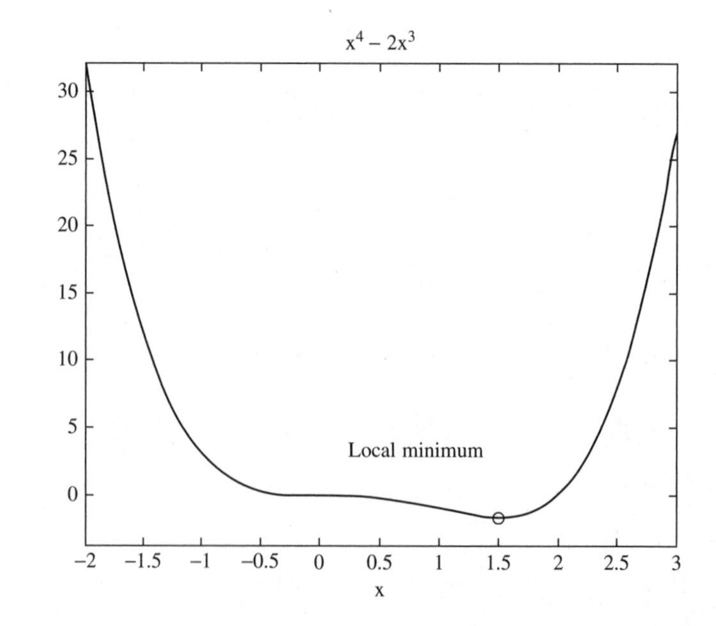

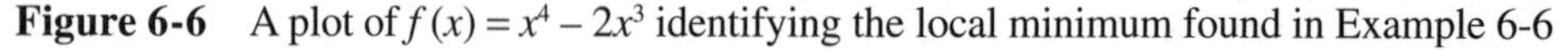

Figure 6-6 A plot of $f(x) = x^4 - 2x^3$ identifying the local minimum found in Example 6-6

The dsolve Command

We can solve differential equations symbolically in MATLAB using the *dsolve* command. The syntax for calling dsolve to find the solution to a single equation is *dsolve('eqn')* where *eqn* is a text string used to enter the equation. This will return a symbolic solution with a set of arbitrary constants that MATLAB labels *C*1, *C*2, and so on. We can also specify initial and boundary conditions for the problem. Conditions are specified as a comma-delimited list following the equation as *dsolve('eqn','cond1', 'cond2',...)* as required.

When using dsolve, derivatives are indicated with a *D*. Hence we enter the equation:

$$\frac{df}{dt} = -2f + \cos t$$

By writing:

```
'Df = -2*f + cos(t)'
```

Higher derivatives are indicated by following D by the order of the derivative. So to enter the equation:

$$y'' + 2y' = 5\sin 7x$$

we would write:

```
'D2y + 2Dy = 5*sin(7*x)'
```

Solving ODE's

Let's begin by considering some trivial differential equations just to get a feel for using MATLAB to deal with ODE's. At the most basic level, we can call dsolve and just pass the equation to it. Here we are with the most basic ODE of all:

```
>>s = dsolve('Dy = a*y')

s =

C1*exp(a*t)
```

Suppose we want to plot the equation for different values of $C1$ and a. We can do this by specifying values for these variables and then assigning them to a new label using the subs command:

```
>> C1 = 2; a = 4;
>> f = subs(s)

f =

2*exp(4*t)
```

Initial conditions are entered in quotes after the equation. For example, suppose that:

$$\frac{dy}{dt} = \frac{t}{t-5} y(t), \; y(0) = 2$$

The call to dsolve in this case is:

```
>> dsolve('Dy = y*t/(t-5)','y(0) = 2')

ans =

-2/3125*exp(t)*(t-5)^5
```

Second and higher order equations work analogously. For example:

$$\frac{d^2y}{dt^2} - y = 0, \quad y(0) = -1, \quad y'(0) = 2$$

can be entered into MATLAB by writing:

```
>> dsolve('D2y - y = 0','y(0) = -1','Dy(0) = 2')

ans =

1/2*exp(t) -3/2*exp(-t)
```

EXAMPLE 6-7

Find a solution to the initial value problem

$$\frac{dy}{dt} = t + 3, \quad y(0) = 7$$

and plot the result for $0 \le t \le 10$.

SOLUTION 6-7

The equation is solved easily with a single call to dsolve:

```
>> s = dsolve('Dy = t + 3','y(0) = 7')

s =

1/2*t^2+3*t+7
```

Now we add a call to ezplot to generate the plot:

```
>> ezplot(s,[0 10])
```

The result is shown in Figure 6-7.

EXAMPLE 6-8

Find the solution of:

$$\frac{dy}{dt} = y^2, \quad y(0) = 1$$

Plot the result showing any asymptotes.

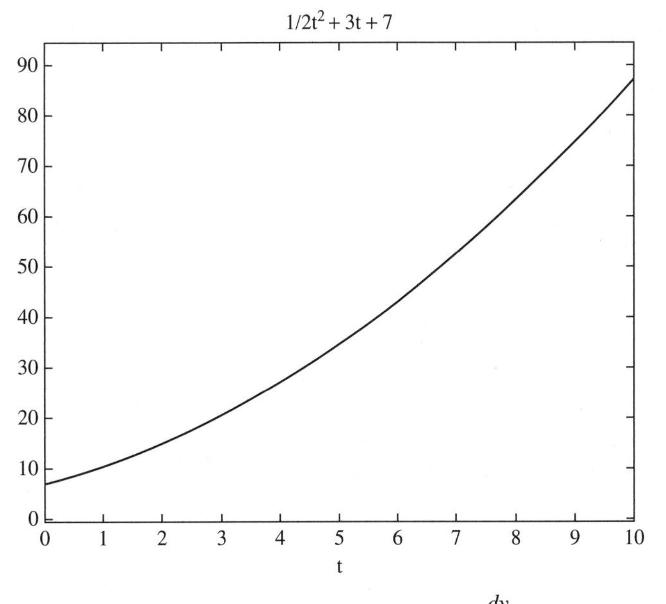

Figure 6-7 A plot of the solution of the IVP $\frac{dy}{dt} = t + 3, y(0) = 7$

SOLUTION 6-8

Using MATLAB dsolve will do the dirty work for us. We find the solution is:

```
>> s = dsolve('Dy = y^2','y(0) = 1')

s =

-1/(t-1)
```

The asymptote is located at:

```
>> d = -1/s

d =

t-1


>> roots = solve(d)

roots =

1
```

Now let's plot and hold it:

```
>> ezplot(s)
>> hold on
```

Now we plot the asymptote:

```
>> plot(double(roots)*[1 1], [-2 2],'--')
>> hold off
```

The result is shown in Figure 6-8.

EXAMPLE 6-9

Solve the BVP given by:

$$\frac{d^2 f}{dx^2} - \frac{\sin x}{x}\left(1 - \frac{2}{x^2}\right) - \frac{2\cos x}{x^2} = 0, \quad f(0) = 2, \quad f'(0) = 0.$$

Plot the solution for $-50 < x < 50$.

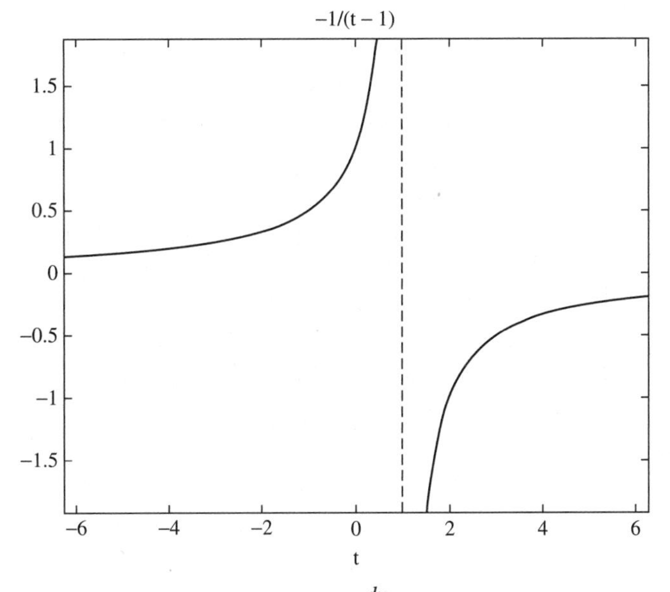

Figure 6-8 Plot of the solution of $\frac{dy}{dt} = y^2, y(0) = 1$ with asymptote

SOLUTION 6-9

By default, dsolve uses t as the independent variable. We can tell it to use something else by tacking the independent variable on at the end of the command. The call to dsolve is:

```
>> g = dsolve('D2f - sin(x)/x-2*cos(x)/x^2+2*sin(x)/x^3 =
0','f(0)=2','Df(0)=0','x')
```

Our declaration of the independent variable is specified at the end following the boundary conditions. The solution returned is:

```
g =

-sin(x)/x+3
```

We can plot the function as shown in Figure 6-9.

EXAMPLE 6-10

Find the general solution of:

$$\frac{dy}{dt} = -\frac{y}{\sqrt{1-t^2}}$$

and plot over $-1 < t < 1$ for the constant $C1 = 0, 10, 20, 30$ on the same graph.

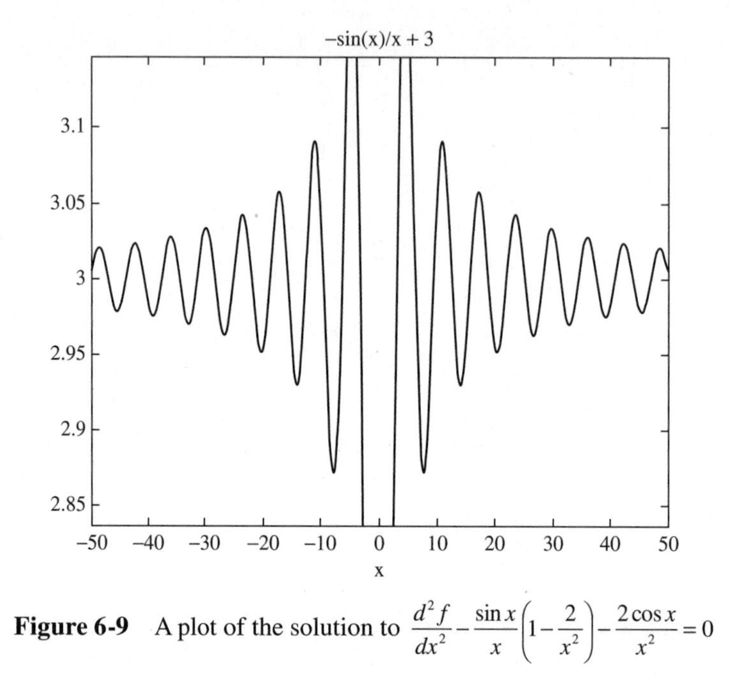

Figure 6-9 A plot of the solution to $\frac{d^2 f}{dx^2} - \frac{\sin x}{x}\left(1 - \frac{2}{x^2}\right) - \frac{2\cos x}{x^2} = 0$

SOLUTION 6-10

We readily obtain the solution using dsolve:

```
>> s = dsolve('Dy = -y/sqrt(1-t^2)')

s =

C1*exp(-asin(t))
```

We can generate multiple curves for different values of C1 on the same graph using a for loop. We create a loop index i and have it assume the values 0, 10, 20, 30. Then we use the subs command to substitute the current value of i and plot the result. The for loop looks like this:

```
>> for i=0:10:30
f = subs(s,'C1',i);
ezplot(f,[-1,1])
hold on
end
```

The first line sets our loop variable to $i = 0, 10, 20, 30$ using the syntax i=start: increment:finish. Next we use subs to tell MATLAB to replace C1 by the current value of the loop variable:

```
f = subs(s,'C1',i);
```

Then we use ezplot to put the curve on the graph. By adding the statement hold on, we tell MATLAB to plot to the same figure each time through the loop. So we end up plotting the functions

```
0, 10*exp(-asin(t)), 20*exp(-asin(t)), 30*exp(-asin(t))
```

The end statement marks the end of the for loop. Next we call hold off so that we close plotting to the current figure and give the plot a title:

```
>> hold off
>> title('IVP Solutions')
```

The result is shown in Figure 6-10.

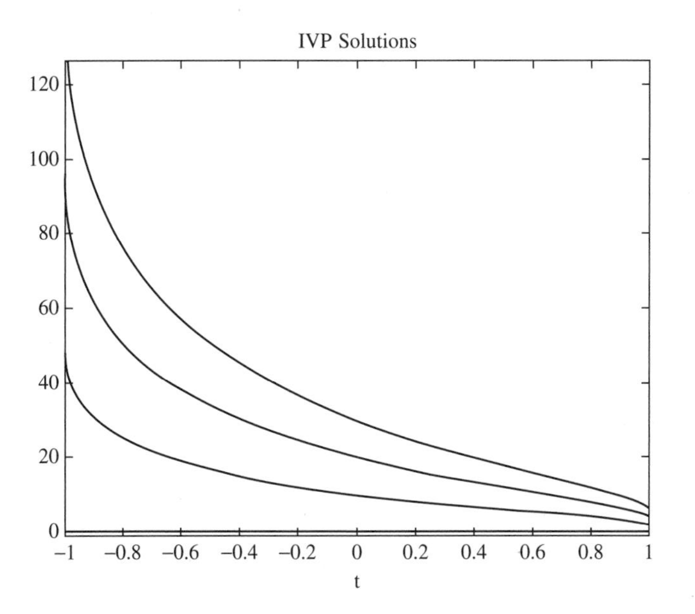

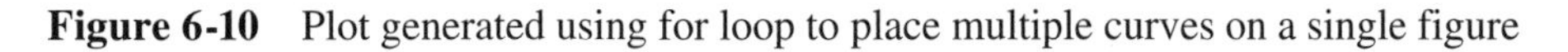
Figure 6-10 Plot generated using for loop to place multiple curves on a single figure

EXAMPLE 6-11

Find a solution of

$$\frac{dy}{dt} = -2ty^2$$

Plot the solution for different initial values, letting $y(0) = 0.2,...,2.0$ in increments of 0.2.

SOLUTION 6-11

First we solve the equation:

```
>> f = dsolve('Dy=-2*t*y^2','y(0)=y0')

f =

1/(t^2+1/y0)
```

We have told MATLAB to set the initial value to a symbol we denoted $y0$. Now we can write a for loop to substitute the values $y(0) = 0.2,\ldots,2.0$. First we define our for loop and loop variable, specifying the start, increment, and end point:

```
for i = 0.2:0.2:2
```

Now we tell MATLAB to substitute i for $y0$ in the solution:

```
temp = subs(f,'y0',i);
```

Next we plot it:

```
ezplot(temp)
```

Finally, we close out the loop by telling MATLAB to hold on so we can add a curve to the same figure each time through the loop, then we end it:

```
hold on
end
```

After the loop runs, we can use the axis command to set the range over each axis to a desired value:

```
>> axis([-4 4 0 2.5])
>> hold off
```

Don't forget to call hold off so that MATLAB will stop sending data to the same figure. The result of all this is show in Figure 6-11.

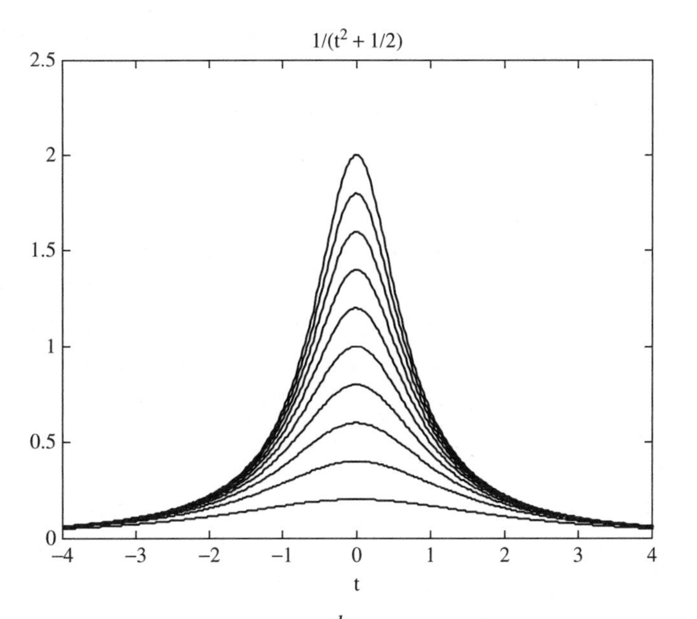

Figure 6-11 A plot of the solution to $\frac{dy}{dt} = -2ty^2$ with initial conditions given by $y(0) = 0.2, \ldots, 2.0$ in increments of 0.2

Systems of Equations and Phase Plane Plots

In this section we will consider the equation for a mass spring system and see how to generate a phase plane plot for the solution. First, how can we use MATLAB to generate a solution to a system of differential equations? The answer is we pass each equation to dsolve.

Consider the simple system:

$$\frac{dX}{dt} = Y, \frac{dY}{dt} = -X$$

$$X(0) = -1, Y(0) = 2$$

The command to enter the system and solve it is:

```
>> s = dsolve('DX = Y','DY = -X','X(0)= -1','Y(0)=2');
```

The solution is returned as a vector. We can extract the solutions by writing:

```
>> s.X
```

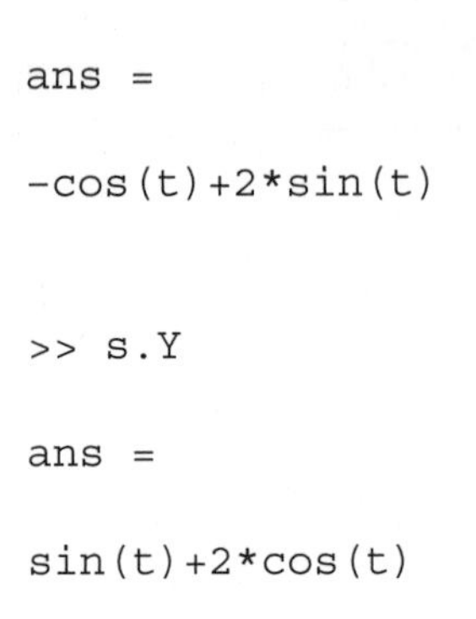

```
ans =

-cos(t)+2*sin(t)


>> s.Y

ans =

sin(t)+2*cos(t)
```

We can show both solutions on one plot:

```
>> ezplot(s.X)
>> hold on
>> ezplot(s.Y)
>> hold off
```

The result is shown in Figure 6-12. This isn't too useful since we can't determine which curve is which.

Let's add some additional commands to clarify the curves in the plot. Suppose that we wanted to plot the solution $X(t)$ using a solid red line and the solution $Y(t)$ using a dashed blue line. How can we do it using ezplot?

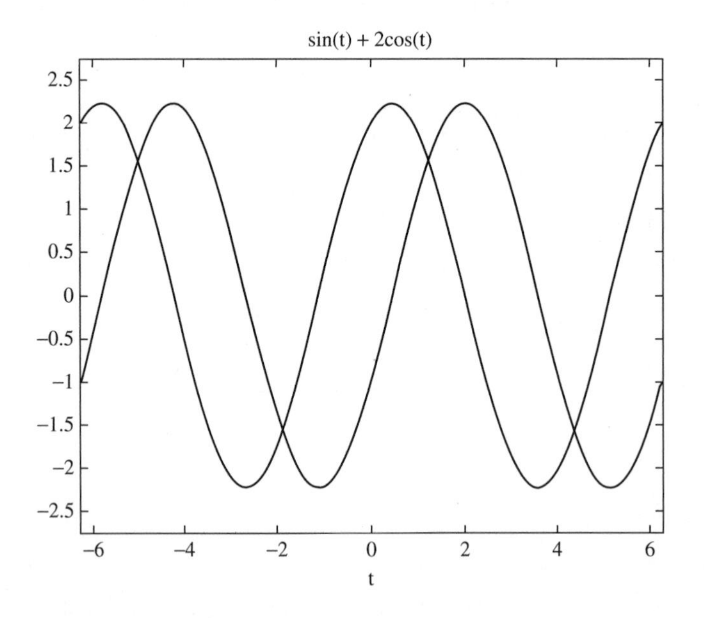

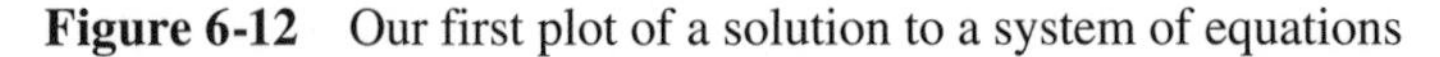

Figure 6-12 Our first plot of a solution to a system of equations

The first step is easy. We can use set to tell MATLAB to draw the first curve with a red line. We can tell MATLAB to find the lines in the plot by calling findobj. Then we tell set we want to color the line red:

```
>> ezplot(s.X),set(findobj('Type','line'),'Color','r')
```

If you try this, you will see a red curve generated nicely on the screen. The problem with this command is that it tells MATLAB to color all lines red, if there are multiple lines on the screen. So to leave this line red and make the second one dashed, we will have to try something different. First we tell MATLAB that we are going to plot another curve on the same figure:

```
>> hold on
```

Now we add the second plot:

```
>> ezplot(s.Y)
```

This will add the curve as a blue line. We need to get a reference to the second curve so that we can ask MATLAB to change it. Next we call get which will return a handle to the current graphics object:

```
>> h=get(gca,'children');
```

Now we can change it using the following command:

```
>> set(h(1),'linestyle','--')
```

The result is displayed in Figure 6-13.

Now let's consider a mass-spring system and see how to obtain solutions to the position and momentum of the mass and generate a phase portrait.

EXAMPLE 6-12

Let $x(t)$ be the position of a mass in a mass-spring system and $p(t)$ be the momentum. Consider the case where the system is governed by a damped oscillator equation:

$$2x'' + x' + 8x = 0,$$

$$\frac{dp}{dt} = -p - 8x$$

$$x(0) = 2, p(0) = x'(0) = 0$$

Find the solutions $x(t)$ and $p(t)$ and generate a phase portrait for the system.

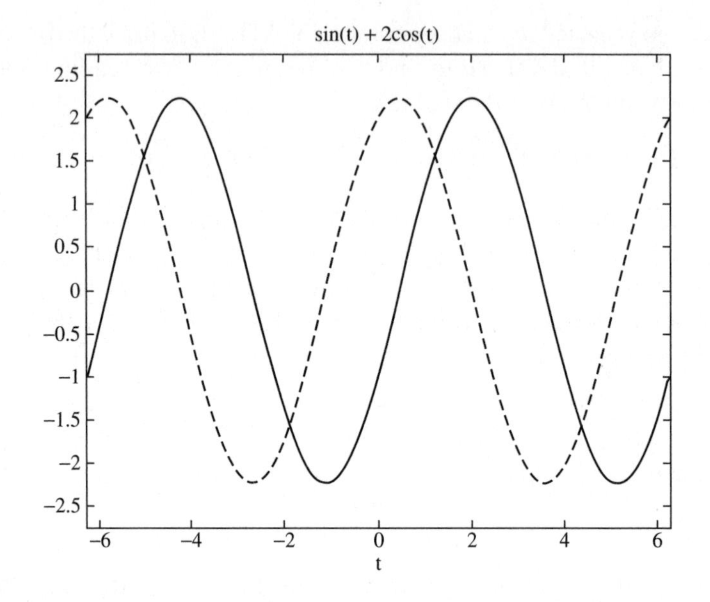

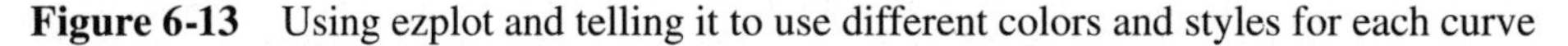
Figure 6-13 Using ezplot and telling it to use different colors and styles for each curve

SOLUTION 6-12

First we call dsolve to generate a solution for the system:

```
>> s = dsolve('2*D2x+Dx+8*x = 0','Dp = -p - 17*x','x(0)=4','
Dx(0)=0','p(0)=0')
```

MATLAB gives a result of the form:

```
s =

    p: [1x1 sym]
    x: [1x1 sym]
```

We can access the solutions by typing s.x and s.p. Let's see what the position is as a function of time for this system:

```
>> s.x

ans =

exp(-1/4*t)*(4/21*sin(3/4*7^(1/2)*t)*7^(1/2)+4*cos(3/4*7^
(1/2)*t))
```

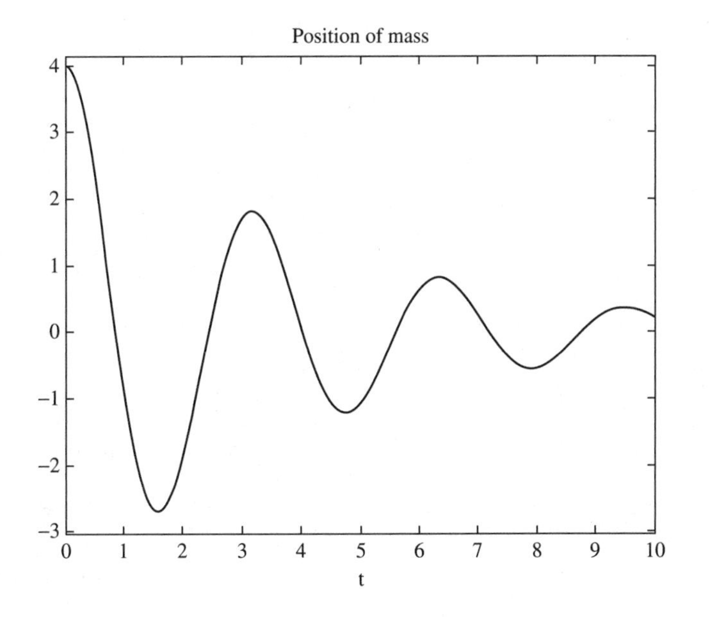

Figure 6-14 The decaying oscillator described by $2x'' + x' + 8x = 0$

For the momentum, we find:

```
>> s.p

ans =

-68/9*cos(3/4*7^(1/2)*t)*exp(-1/4*t)-748/63*7^(1/2)*exp(-
1/4*t)*sin(3/4*7^(1/2)*t)+68/9*exp(-t)
```

Now let's plot the position as a function of time. We use:

```
>> ezplot(s.x, [0 10])
>> title('Position of Mass')
```

As can be seen in Figure 6-14, the position of the mass is described by an oscillator that decays exponentially, as expected for a damped system.

Now let's plot the momentum over the same time range:

```
>> ezplot(s.p, [0 10])
>> title('Momentum')
```

As shown in Figure 6-15, the momentum is also a decaying oscillator, but the amplitude is much larger.

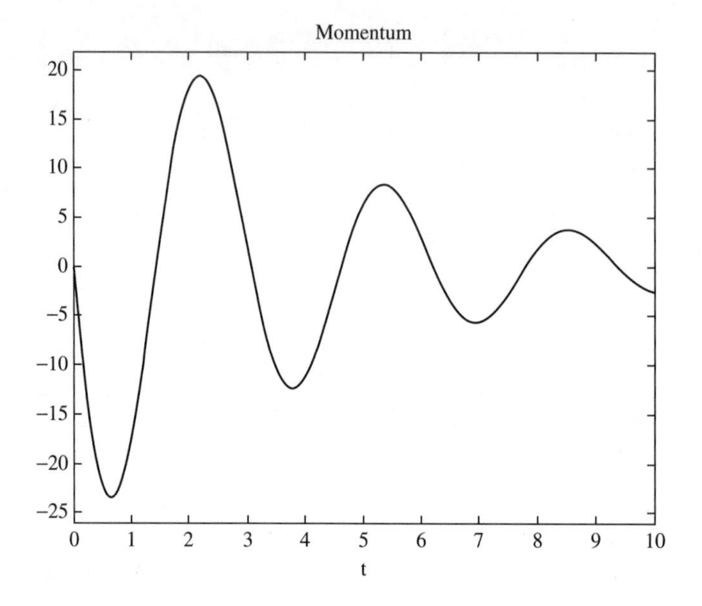

Figure 6-15 The momentum of the system with $2x'' + x' + 8x = 0$

A phase plot is a plot of p versus x. We can try this using ezplot in the following way. We call it passing both functions at the same time:

```
>> ezplot(s.x,s.p,[-5 5])
```

We fix up the axes a little bit and generate a title:

```
>> axis([-8 8 -25 20])
>> title('parametric plot')
```

The result is displayed in Figure 6-16.

You might think the plot looks a bit funny, so let's try something else. We can also generate a parametric plot numerically. First we define a numerically generated time interval:

```
>> tvalues = (0:0.1:10);
```

Now we use subs to generate numerical representations of the position and momentum functions over this time interval:

```
>> xval = subs(s.x,'t',tvalues);
>> pval = subs(s.p,'t',tvalues);
```

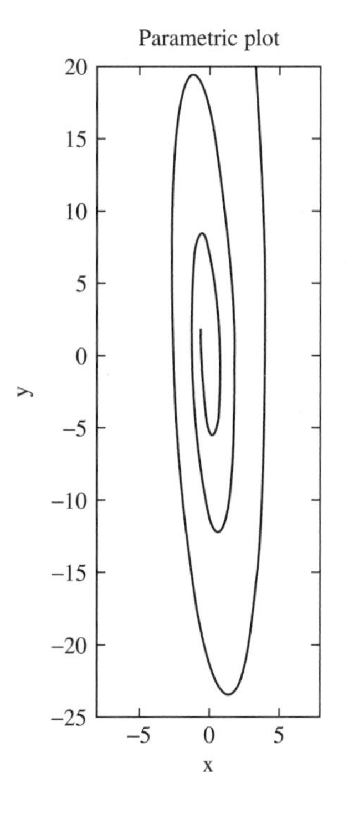

Figure 6-16 A phase plot generated using the ezplot function

Now let's generate the phase portrait by calling plot in this way:

```
>> plot(xval,pval),xlabel('x'),ylabel('p'),title
('phase portrait for mass-spring')
```

The result, which is much nicer, is shown in Figure 6-17.

EXAMPLE 6-13
Generate a solution of the critically damped system:

$$\frac{d^2x}{dt^2}+\frac{dx}{dt}+\frac{1}{4}x=0, \quad \frac{dp}{dt}=-\frac{1}{2}p-\frac{1}{4}x$$

$$x(0)=4, \quad p(0)=0$$

SOLUTION 6-13
We enter the equations and solve:

```
>> s = dsolve('D2x + Dx + (1/4)*x = 0','Dp + (1/2)*p +
(1/4)*x = 0','x(0) = 4','Dx(0)=0','p(0)=0');
```

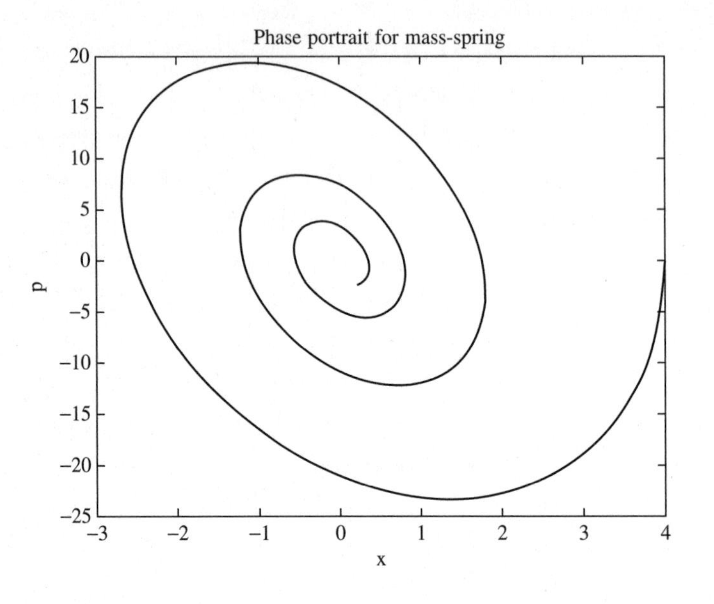

Figure 6-17 A numerically generated phase portrait

The results are:

```
>> s.x

ans =

exp(-1/2*t)*(4+2*t)


>> s.p

ans =

-1/8*(8*t+2*t^2)*exp(-1/2*t)
```

Let's plot the position as a function of time, which is shown in Figure 6-18:

```
>> ezplot(s.x,[0 15])
>> axis([0 15 0 4])
>> title('Position')
```

Now let's generate a plot of the momentum:

```
>> ezplot(s.p,[0 15])
>> title('Momentum')
```

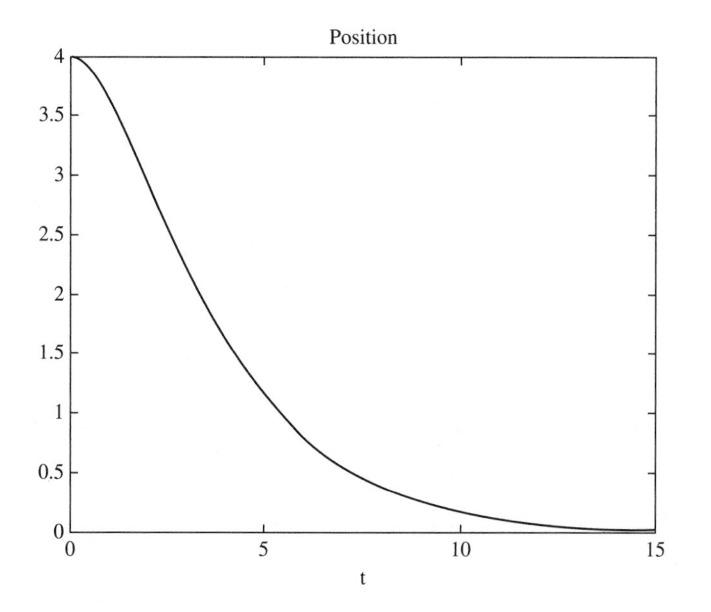

Figure 6-18 Position as a function of time for a critically damped oscillator

The momentum plot is shown in Figure 6-19. Notice that the momentum and position of the mass quickly damp out with no oscillations, as expected for a critically damped case.

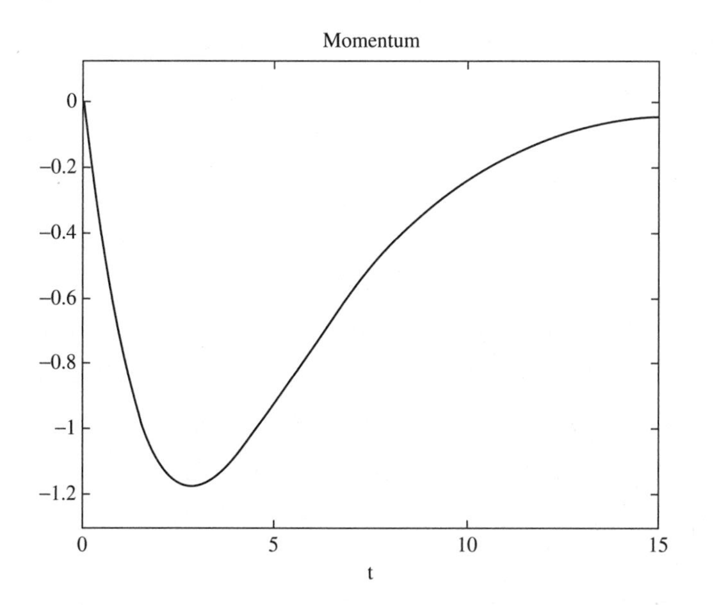

Figure 6-19 The momentum for a critically damped oscillator

Quiz

1. Calculate $\lim_{x\to 0}\frac{\sin 4x}{x}$.
2. Find $\lim_{x\to 2}\frac{x-2}{|x-2|}$. Does it exist?
3. Find the asymptotes of $\frac{x}{x^2-3x}$ and plot them as dashed lines with the function.
4. Compute $\lim_{\theta\to\pi/2}\frac{\cos\theta}{1+\sin\theta}$.
5. Consider the function $f(x)=\frac{1}{3x^2+1}$.

   a) What are the first and second derivatives?

   b) What are the critical points for this function?

   c) What is f'' evaluated at these critical points?

   d) Are there any local minima or maxima? Plot the function and show them on the graph.
6. Use MATLAB to show that $y = 2 \sin t - 3 \cos t$ does not satisfy $y'' - 11y = -4 \cos 6t$.
7. Find a solution of $dx/dt = -2x + 8$ and plot for constants $C1 = 1, 2, \ldots 5$ on the same graph.
8. Find a solution of $y'' - y' - 2y = 2t$, $y(0) = 0$, $y'(0) = 0$.
9. Solve the system $x'' + x = 0$, $p' + p + x = 0$, $x(0) = 4$, $x'(0) = p(0) = 0$.
10. Solve the system $x'' + 2x' + x = 0$, $p' = x$, $x(0) = 1$, $x'(0) = p(0) = 0$ and generate a phase portrait.

CHAPTER 7

Numerical Solution of ODEs

In the last chapter we learned how to solve symbolic ordinary differential equations. In this chapter we will learn how to use MATLAB to generate numerical solutions of ODEs. MATLAB has several solvers that can be used to solve differential equations as we will see in the following text.

Solving First Order Equations with ODE23 and ODE45

To solve an ODE numerically, first define a function that can be used to represent the equation. For our first example, we will consider the equation:

$$\frac{dy}{dt} = \cos(t)$$

We can readily integrate the equation giving $y(t) = \sin(t) + C$, so it will be easy for us to check the solution we obtain numerically. First let's define our function. We do this by creating a .m file and typing the following input:

```
function ydot = eq1(t,y)
ydot = cos(t);
```

To solve the ODE, we make a call to the function ODE23. This function works by integrating the differential equations using second and third order Runge-Kutta method. The syntax used is:

```
[t,y] = ode23('func_name', [start_time, end_time], y(0))
```

Our function is called eq1. Let's solve it over the time interval $0 \le t \le 2\pi$ and suppose that $y(0) = 2$. The call would then look like this:

```
>> [t,y] = ode23('eq1',[0 2*pi],2);
```

Since this is a simple equation that can be solved readily, MATLAB soon returns with the answer. Since we are doing numerical work, to see what the solution is we need to plot it. First let's generate a data set representing the analytical solution so we can compare:

```
>> f = 2 + sin(t);
```

Now we use the following command to generate the plot:

```
>> plot(t,y,'o',t,f),xlabel('t'),ylabel('y(t)'),axis([0 2*pi 0 4])
```

The plot is shown in Figure 7-1. The circles mark the solution returned by ode23, while the solid line represents the analytical solution. It looks like the solution returned by ode23 is pretty close.

How close is the solution? We can check it by calculating the relative error. If $f(t)$ represents the analytical solution and $y(t)$ represents the numerical solution, then the relative error is given by:

$$\left| \frac{f(t) - y(t)}{f(t)} \right|$$

We will call our error function err. First let's allocate memory for an array to store the error at each point. We do this by creating an array of zeros:

```
>> err = zeros(size(y));
```

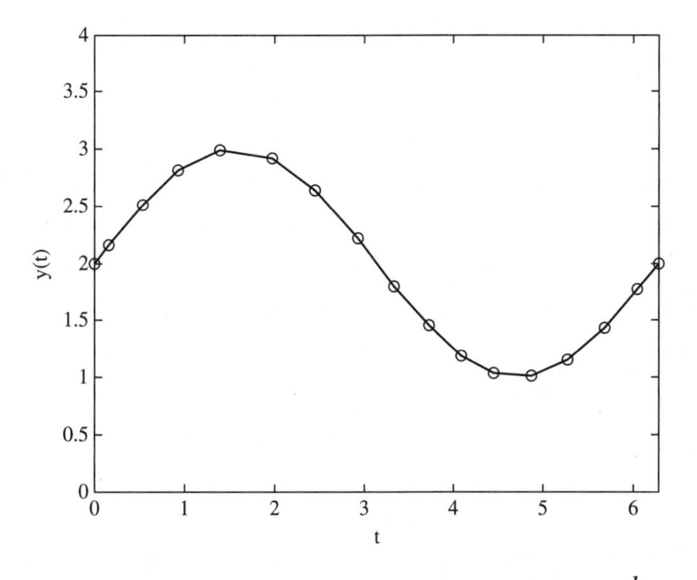

Figure 7-1 Plot showing numerical and analytical solutions of $\frac{dy}{dt} = \cos(t)$

Now we calculate the relative error at each point by using a for loop to move through the data:

```
>> for i = 1:1:size(y)
err(i) = abs((f(i)-y(i))/f(i));
end
```

Let's look at the result:

```
>> err

err =

  1.0e-003 *

         0
    0.0001
    0.0091
    0.0301
    0.0743
    0.2115
    0.2876
    0.3780
    0.4659
```

```
    0.5597
    0.6480
    0.6907
    0.6050
    0.4533
    0.3164
    0.2414
    0.2129
```

The errors are quite small in this case. The largest error is:

```
>> emax = max(err)

emax =

  6.9075e-004
```

Let's look at the data points returned by the ode23 solver:

```
y =

    2.0000
    2.1593
    2.5110
    2.8080
    2.9848
    2.9169
    2.6344
    2.2123
    1.7999
    1.4512
    1.1892
    1.0323
    1.0115
    1.1536
    1.4337
    1.7689
    1.9996
```

With the precision used to present these numbers, since the relative error is much smaller, we can be satisfied with the results.

The ode45 function uses higher order Runge-Kutta formulas. Let's see how it works differently in the case we are currently working with. The call is similar:

```
>> [t,w] = ode45('eq1',[0 2*pi],2);
```

However, the solution is evaluated over a larger number of points. So we need to regenerate the analytical solution, then let's plot them both:

```
>> f = 2 + sin(t);
>> plot(t,w,'o',t,f),xlabel('t'),ylabel('y(t)'),axis([0 2*pi 0 4])
```

The result is shown in Figure 7-2. You can see that the density of points from the numerical solution is higher than what we got with ode23.

Let's compare the sizes of the solutions:

```
>> size(w),size(y)

ans =

    45    1

ans =

1
```

The ode45 solver has returned 45 data points while the ode23 solver returned 17 data points. Whether this is important or not will depend on your application. Let's create another array of zeros and calculate the relative error:

```
>> err = zeros(size(w));
>> for i = 1:1:size(w)
err(i) = abs((f(i)-w(i))/f(i));
end
```

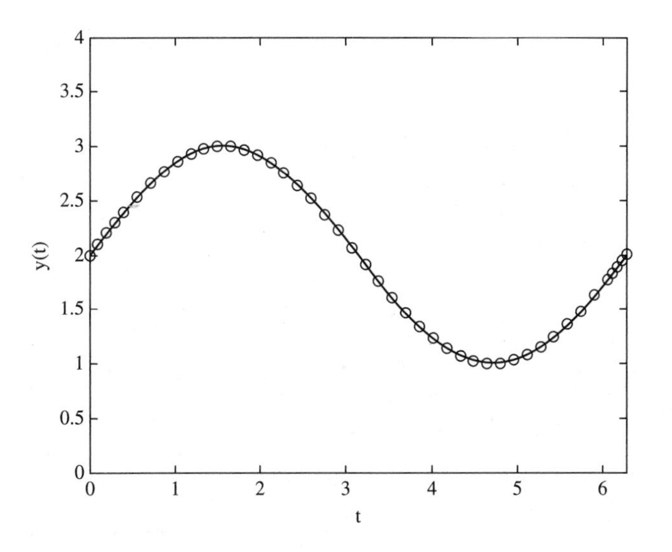

Figure 7-2 Solving the equation again with ode45

Let's find the maximum relative error this time:

```
>> wmax = max(err)

wmax =

   4.9182e-006
```

If you recall, the maximum error we found with the ode23 solver was:

```
>> emax

emax =

   6.9075e-004
```

This is quite a bit larger than the maximum relative error obtained with ode45. In fact it's almost 141 times as big:

```
>> emax/wmax

ans =

   140.4473
```

Hence it may be wise to use ode45, if the error differences are important. Let's consider another example.

EXAMPLE 7-1

Consider the ordinary differential equation with forcing:

$$\frac{dy}{dt} = 5(F(t) - y)$$

The forcing function is:

$$F(t) = te^{-t/tc}\cos(\omega t)$$

Find and plot the numerical solution for the following values of the time constant *tc* and frequency ω shown in Table 7-1.

Table 7-1 Time constant and frequency values for Example 7-1.

Time Constant	Frequency
0.01 s	628 rad/s
0.1 s	6.28 rad/s

SOLUTION 7-1

To facilitate the ability to use several different values of the time constant and frequency, we will declare these as global values. First let's write a function to implement the equation:

```
function ydot = eq2(t,y)
global tc w
F = t*exp(-t/tc)*cos(w*t);
ydot = 5*F;
```

Following the procedure we used in the text, we code this function up in a .m file and save it. First we define our global variables (the time constant and frequency) in MATLAB:

```
>> global tc w
```

Now let's assign the first values in the table:

```
>> tc = 0.01; w = 628;
```

Now let's set the end time of the interval we will evaluate together with the initial condition:

```
>> final_time = 0.1; y0 = 0;
```

Next we call ode45 and generate the first solution:

```
>> [t,y] = ode45('eq2',[0 final_time],y0);
```

Let's plot the solution in Figure 7-3.
Let's compare it with the forcing function:

```
>> plot(t,y,t,F,'--'),xlabel('t')
```

The forcing function is the dashed line. As you can see in Figure 7-4, it is quite a bit larger than the response—but it has a similar functional form. In both cases we see an exponentially decaying oscillator.

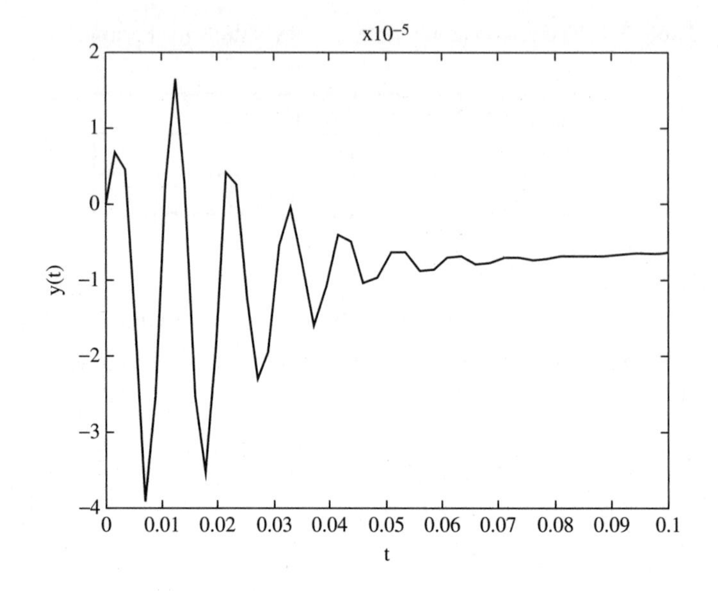

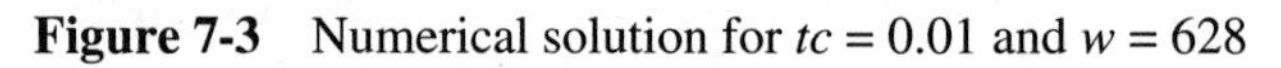
Figure 7-3 Numerical solution for *tc* = 0.01 and *w* = 628

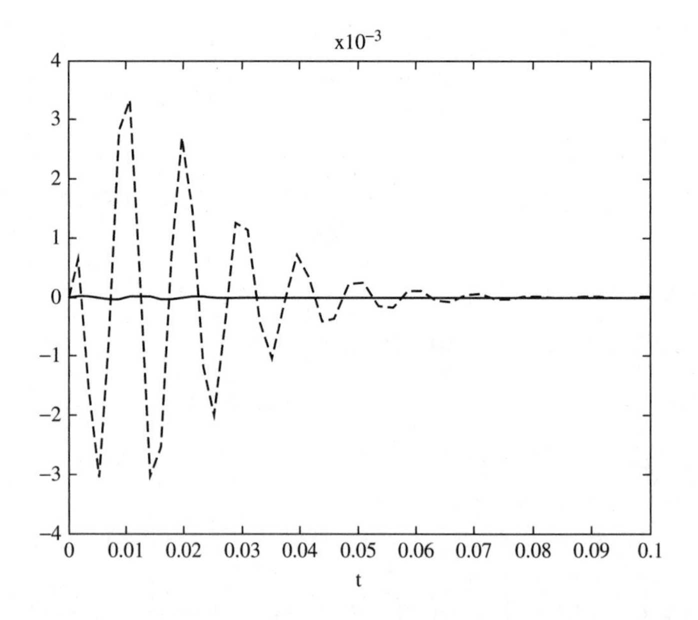

Figure 7-4 A comparison of the forcing function to the response. In this case, the forcing seems to continue far longer than the response

Now let's try the next set of values.

```
>> tc = 0.1; w = 6.28;
```

Since the time constant is much larger, let's increase our time interval:

```
>> final_time = 1.0;
```

Now we call solver:

```
>> [t,y] = ode45('eq2',[0 final_time],y0);

>> plot(t,y),xlabel('t'),ylabel('y(t)'),axis([0 0.7 0 0.015])
```

The plot in this case is shown in Figure 7-5.

Let's plot the forcing function and response together on the same plot. In this case, you can see the response clearly lagging the forcing function. This is shown in Figure 7-6. The command to plot showing the forcing function as a dashed line is:

```
>> plot(t,y,t,F,'--'),xlabel('t'),ylabel('y(t)'),axis([0 0.7 0 0.05])
```

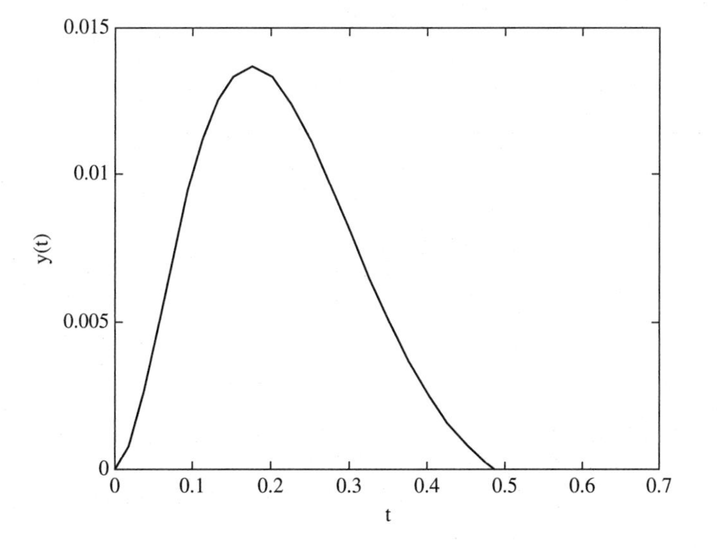

Figure 7-5 The response for $\frac{dy}{dt} = 5(F(t) - y)$ when $\omega = 6.28$ and $tc = 0.1$

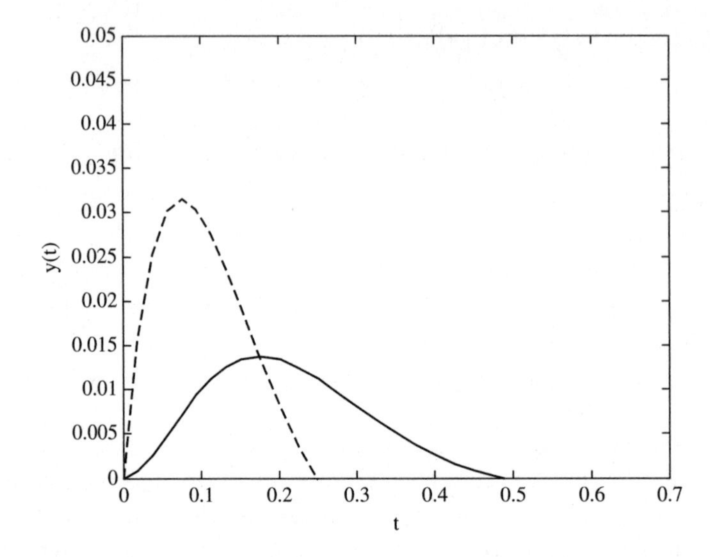

Figure 7-6 In the second case examined in Example 7-1, we get a nice plot showing the response lagging the forcing function, with both functions having the same general shape

Solving Second Order Equations

Now let's see how to solve second order equations numerically. The trick used here is to break the equation down into a system of two equations. So first, let's see how to solve a system of equations.

EXAMPLE 7-2
Solve the system

$$\frac{dx}{dt} = -x^2 + y, \frac{dy}{dt} = -x - xy$$

with initial conditions $x(0) = 0$, $y(0) = 1$. Plot both functions and generate a phase plot.

SOLUTION 7-2
First we create a function like before that will implement the right-hand side of the differential equation. This time, of course, we have a system so need a 2×1 column vector. The function looks like this:

```
function xdot = eqx(t,x);
%allocate array to store data
```

```
xdot = zeros(2,1);
xdot(1) = -x(1)^2 + x(2);
xdot(2) = -x(1) - x(1)* x(2);
```

So far so good, pretty simple. Now let's use ode45 to generate a solution with the given initial conditions:

```
>> [t,x] = ode45('eqx',[0 10],[0,1]);
```

Now let's generate a plot. We access the first function with x(:,1) and the second by writing x(:,2). The plot command is:

```
>> plot(t,x(:,1),t,x(:,2),'--'),xlabel('t'), axis([0 10 -1.12 1.12])
```

The plot of the two functions is shown in Figure 7-7.

Now let's generate the phase plot. The command is:

```
>> plot(x(:,1),x(:,2))
```

The phase plot is shown in Figure 7-8.

Now that we know how to solve a system, we can use ode45 to solve a second order equation.

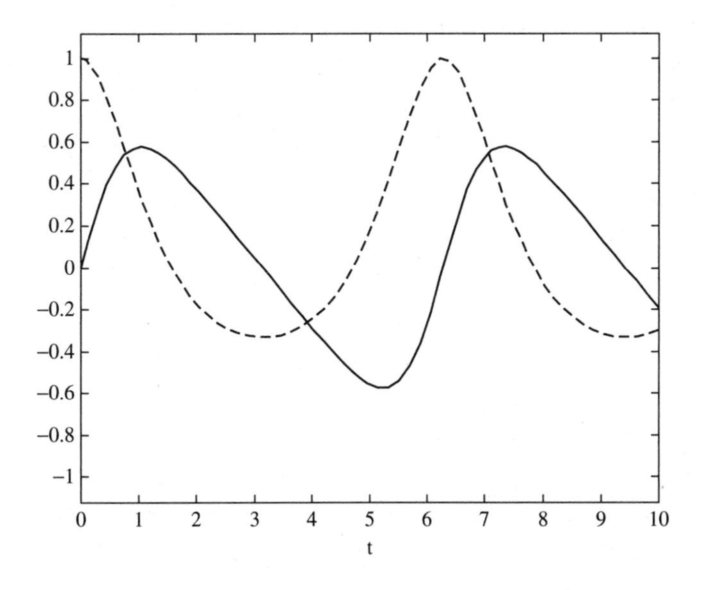

Figure 7-7 The functions *x* (solid line) and *y* (dashed line) that solve the system in Example 7-2

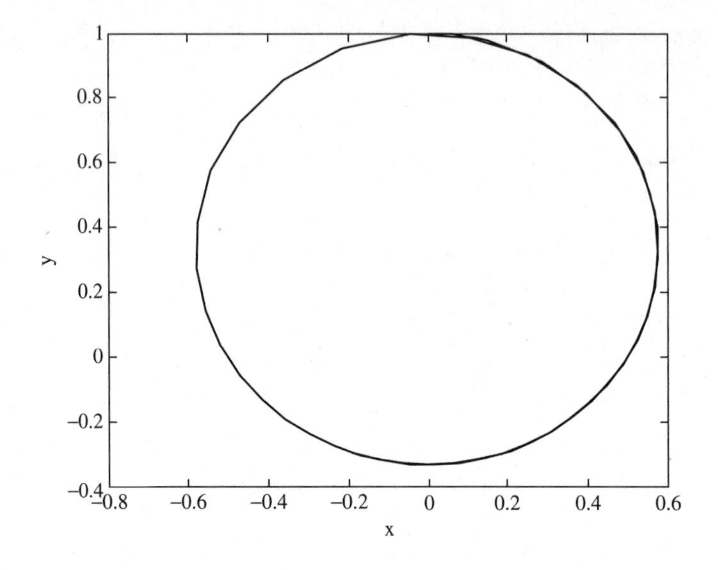

Figure 7-8 The phase portrait for the system solved in Example 7-2

EXAMPLE 7-3

Find a solution of $y'' + 16y = \sin(4.3t)$ when $y(0) = y'(0) = 0$.

SOLUTION 7-3

We can change this into a system of first order equations. First we set:

$$x_1 = y$$
$$x_2 = y'$$

Hence:

$$x_1' = y' = x_2$$
$$x_2' = y'' = \sin(4.3t) - 16x_1$$

Now we create a function to implement this system:

```
function xdot = eqx2(t,x);
%allocate array to store data
xdot = zeros(2,1);
xdot(1) = x(2);
xdot(2) = sin(4.3*t)-16*x(1);
```

Let's call ode45 to get a solution. Since the forcing function is sinusoidal, we choose a time interval $0 \le t \le 2\pi$:

```
>> [t,x] = ode45('eqx2',[0 2*pi],[0,0]);
```

Now let's plot the functions that are returned.

```
>> plot(t,x(:,1),t,x(:,2),'--'),xlabel('t'), axis([0 2*pi -3 3])
```

The plot is shown in Figure 7-9.

Notice that the amplitudes of the solutions are growing with time. They are also out of phase and the second solution has a much larger amplitude than the first. Let's generate a phase plot for the system:

```
>> plot(x(:,1),x(:,2)),xlabel('x1'),ylabel('x2')
```

The result is shown in Figure 7-10.

For comparison, let's consider the case where the initial conditions are given by $y(0) = y'(0) = 1$. Let's redo the solution and plot it:

```
>> [t,x] = ode45('eqx2',[0 2*pi],[1,1]);
>> plot(t,x(:,1),t,x(:,2),'--'),xlabel('t'), axis([0 2*pi -4 4])
```

The plot is shown in Figure 7-11.

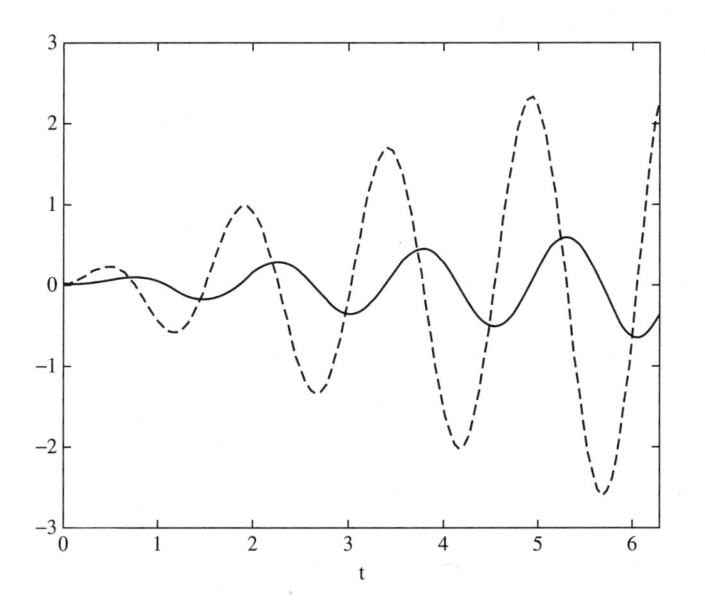

Figure 7-9 The system solved in Example 7-2. Notice it exhibits beats

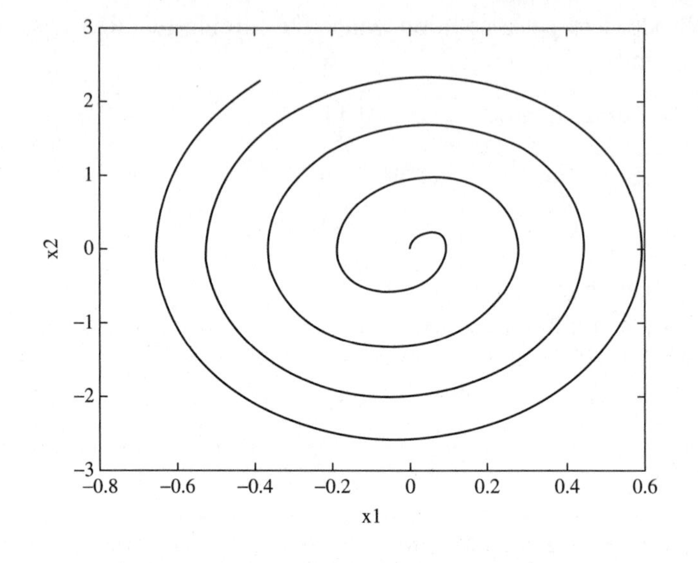

Figure 7-10 A phase portrait for the system in Example 7-2 with initial conditions given by $y(0) = y'(0) = 0$

Notice that by changing the initial conditions, we have increased the amplitude of the oscillations by a large fraction—the functions get bigger earlier. The behavior appears more regular for x_1. Let's take a closer look at this and compare x_1 with $\cos(4.3t)$ in a single plot, shown in Figure 7-12.

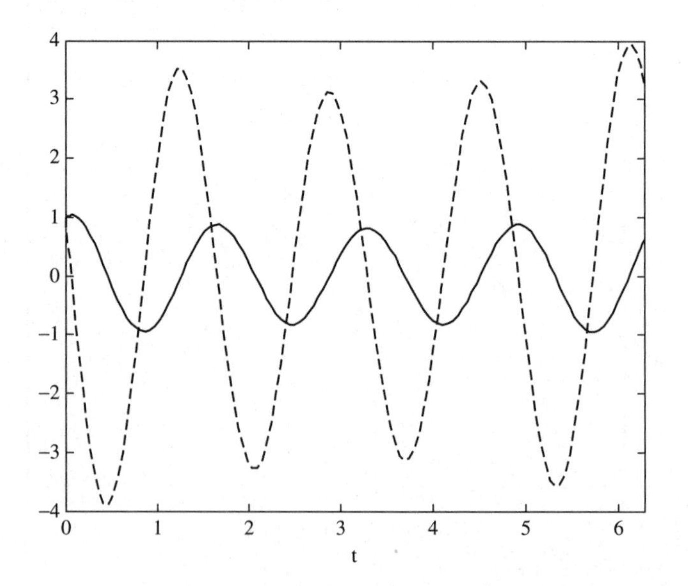

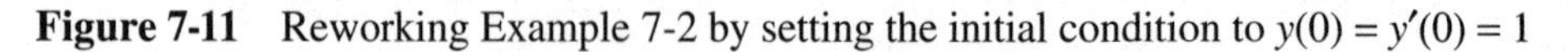

Figure 7-11 Reworking Example 7-2 by setting the initial condition to $y(0) = y'(0) = 1$

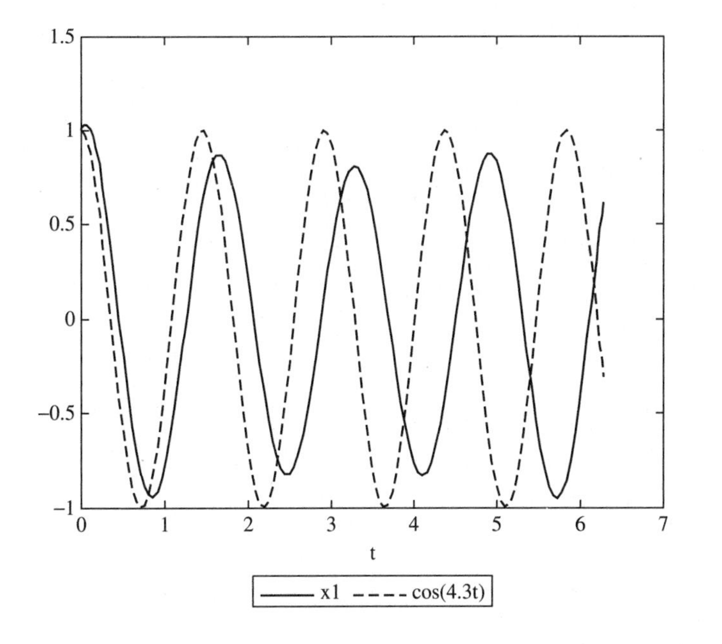

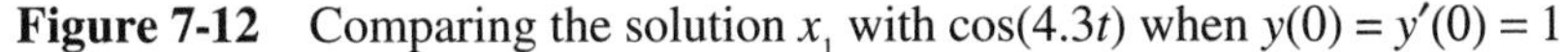

Figure 7-12 Comparing the solution x_1 with cos(4.3t) when $y(0) = y'(0) = 1$

The two functions start off pretty close, but diverge as time goes on. But compare this to the first case, when $y(0) = y'(0) = 0$. In that case the solution had little resemblance to something we know well as shown in Figure 7-13.

Let's take a look at the beating phenomenon by solving over a larger time interval:

```
>> [t,x] = ode45('eqx2',[4*pi 20*pi],[0,0]);
>> plot(t,x(:,1)),xlabel('t')
```

The result is shown in Figure 7-14, where you can see the familiar form of a system with beats.

Let's return to the case where $y(0) = y'(0) = 1$. In Figure 7-15, we show the plot of the solution over the same time interval shown in Figure 7-14. Look closely and you will notice that in Figure 7-14 which shows the plot for the case of $y(0) = y'(0) = 0$, there is a phase reversal each time the system goes down near zero oscillation and starts to build up again. This feature is not present when $y(0) = y'(0) = 1$.

Finally, we generate the phase plot for the case of $y(0) = y'(0) = 1$. This is shown in Figure 7-12. Notice the difference as compared to Figure 7-10. What does that tell you about the solutions?

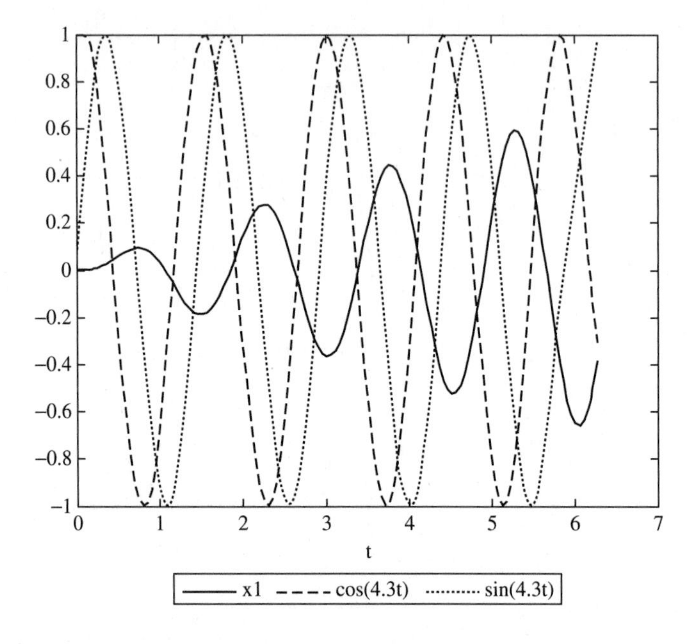

Figure 7-13 The solid line represents the solution when $y(0) = y'(0) = 0$. It bears little resemblance to the forcing trig function or its derivative

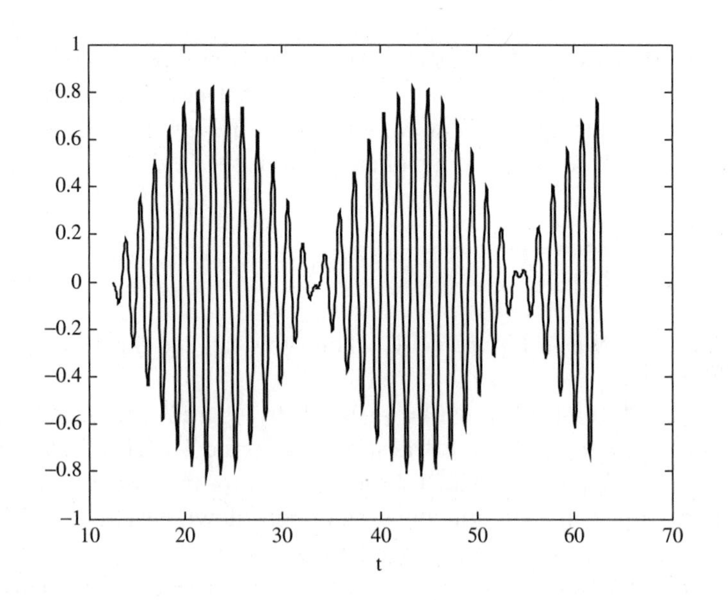

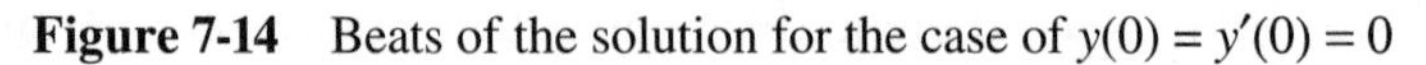

Figure 7-14 Beats of the solution for the case of $y(0) = y'(0) = 0$

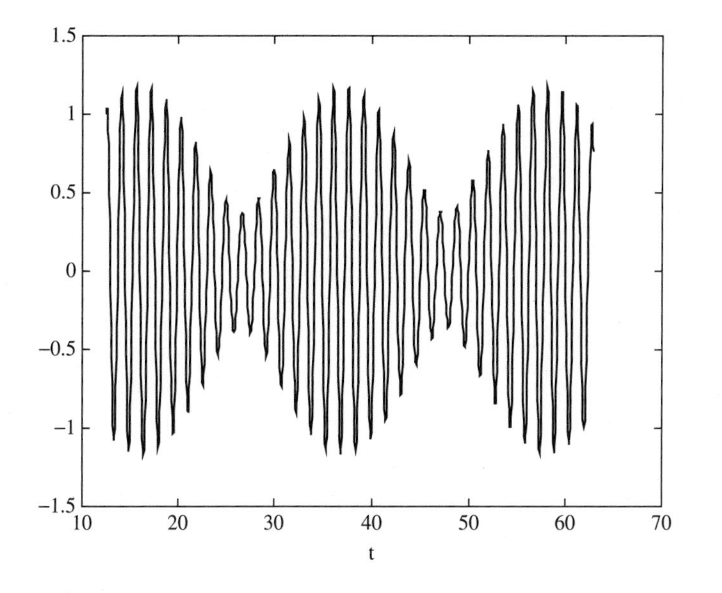

Figure 7-15 When $y(0) = y'(0) = 1$, the amplitude of the solution shrinks and then gets larger again—but there is no phase reversal

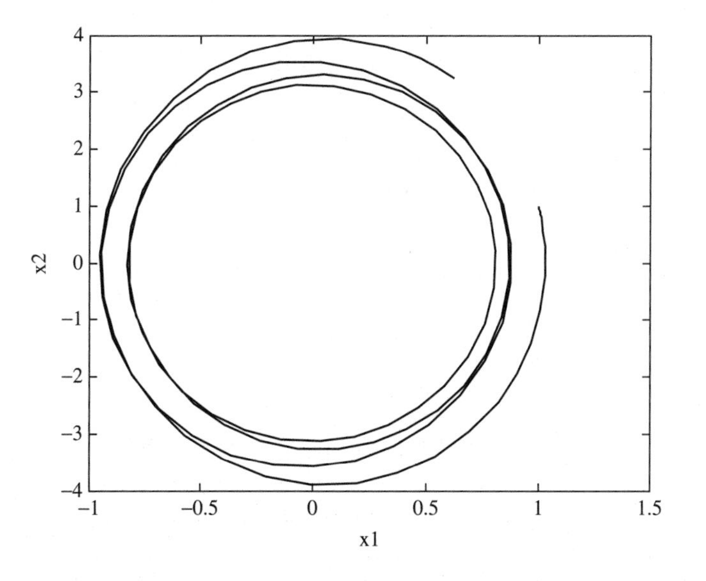

Figure 7-16 Phase portrait for the system in Example 7-2, with $y(0) = y'(0) = 1$

Quiz

1. Find a solution of the system:

$$x_1' = 2x_1x_2$$
$$x_2' = -x_1^2$$
$$x_1(0) = x_2(0) = 0$$

2. Find a solution of the system:

$$x_1' = x_2$$
$$x_2' = -x_1$$
$$x_1(0) = x_2(0) = 1$$

3. Solve:

$$y' = -2.3y, \quad y(0) = 0$$

4. Use ode23 and ode45 to solve:

$$y' = y, \quad y(0) = 1$$

   How many points are returned by ode23 and ode 45?

5. Solve $y' = -ty + 1$, $y(0) = 1$.
6. Solve $y' = t^2y = 0$, $y(0) = 1$ for $0 \le t \le 2$.
7. What numerical technique is used by ode23 and ode45?
8. Solve $\dfrac{dy}{dt} = \dfrac{2}{\sqrt{1-t^2}}, -1 < t < 1, y(0) = 1.$
9. Solve $y'' - 2y' + y = \exp(-t)$, $y(0) = 2$, $y'(0) = 0$.
10. Generate a phase portrait for the system in problem 9.

CHAPTER 8

Integration

This chapter covers methods that can be used to compute integrals. We will begin with integration of symbolic expressions and then consider numerical integration.

The Int Command

Let f be a symbolic expression in MATLAB. We can derive an expression giving the indefinite integral of f by writing:

```
int(f)
```

The expression f can be entered by creating a variable or reference first or by directly passing a quote-delimited string to int. For example, we can show that:

$$\int x \, dx = \frac{1}{2}x^2$$

(leaving out the constant of integration) by writing:

```
>> int('x')
```

MATLAB returns:

```
ans =

1/2*x^2
```

MATLAB can generate integrals that are entirely symbolic. That is, instead of:

```
>> int('x^2')

ans =

1/3*x^3
```

Consider:

```
>> int('x^n')

ans =

x^(n+1)/(n+1)
```

If we don't tell it anything, int will make assumptions about what variable to integrate. For example, we can define a trig function:

```
>> g = 'sin(n*t)';
```

If we just pass this function to int, it assumes that t is the integration variable:

```
>> int(g)

ans =

-1/n*cos(n*t)
```

However we can call int using the syntax int(f, v) where f is the function to integrate and v is the integration variable. Using g again we could write:

```
>> syms n
>> int(g,n)

ans =

-1/t*cos(n*t)
```

For readers taking calculus, don't forget to add the constant of integration to your answer. When creating symbolic expressions, it's not always necessary to enter them in quotes—remember to use the syms command. If we type:

```
>> g = b^t;
```

MATLAB complains, saying:

```
??? Undefined function or variable 't'.
```

We can get around this in the following way. First we call syms and tell MATLAB what we want to use for symbolic variables, and then we can define our functions without enclosing them in quotes:

```
>> syms a t
>> g = a*cos(pi*t)

g =

a*cos(pi*t)


>> int(g)

ans =

a/pi*sin(pi*t)
```

EXAMPLE 8-1

What is the integral of $f(x) = b^x$? Evaluate the resulting expression for $b = 2$, $x = 4$.

SOLUTION 8-1

We start by defining our symbolic variables:

```
>> syms b x
```

Now we define the function and integrate:

```
>> f = b^x;
>> F = int(f)

F =

1/log(b)*b^x
```

We can obtain a numerical value for the expression with the given values by calling subs. To substitute numerical values for multiple symbolic variables in one command, enclose the variable list and substitution list in curly braces { }. In this case we write:

```
>> subs(F,{b,x},{2,4})

ans =

   23.0831
```

EXAMPLE 8-2
Compute $\int x^5 \cos(9x)\,dx$.

SOLUTION 8-2
Doing this integral by hand would require integration by parts and a great deal of pain. With MATLAB, we can generate the answer on a single line:

```
>> F = int(x^5*cos(x))

F =

x^5*sin(x)+5*x^4*cos(x)-20*x^3*sin(x)-60*x^2*cos(x)+120*cos(x)+120*x*sin(x)
```

We can use the command "pretty" to have MATLAB display the answer in a more pleasing format:

```
>> pretty(F)
```

$$x^5 \sin(x) + 5x^4 \cos(x) - 20x^2 - 60x \cos(x) + 120 \cos(x) + 120x \sin(x)$$

EXAMPLE 8-3
Find $\int 3y \sec(x)\,dy$.

SOLUTION 8-3
The integrand contains two variables, so we tell MATLAB that we want to integrate with respect to y:

```
>> int('3*y^2*sec(x)',y)

ans =

y^3*sec(x)
```

Had we wanted to integrate with respect to x instead we would have written:

```
>> int('3*y^2*sec(x)',x)

ans =

3*y^2*log(sec(x)+tan(x))
```

Definite Integration

The int command can be used for definite integration by passing the limits over which you want to calculate the integral. If we enter int(f, a, b) then MATLAB integrates over the default independent variable and returns:

$$\int_a^b f(x)\,dx = F(b) - F(a)$$

For example:

$$\int_2^3 x\,dx = \frac{1}{2}x^2\Big|_2^3 = \frac{1}{2}(9-4) = \frac{5}{2}$$

would be calculated in MATLAB by writing:

```
>> int('x',2,3)

ans =

5/2
```

Equivalently, if we wanted MATLAB to generate the intermediate expression $\frac{1}{2}x^2$, we could write:

```
>> F = int('x')

F =

1/2*x^2
```

```
>> a = subs(F,x,3)-subs(F,x,2)

a =

    2.5000
```

EXAMPLE 8-4

What is the area under the curve $f(x) = x^2 \cos x$ for $-6 \leq x \leq 6$?

SOLUTION 8-4

The curve is shown in Figure 8-1. To find the area under the curve, we compute:

$$\int_{-6}^{6} x^2 \cos x \, dx$$

Let's define the function:

```
>> f = x^2*cos(x);
```

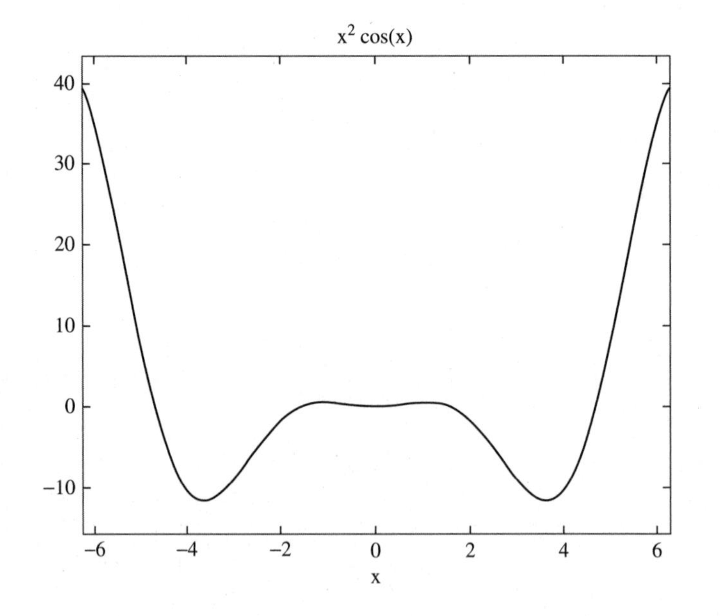

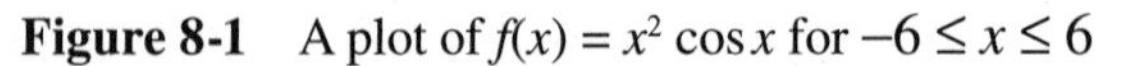

Figure 8-1 A plot of $f(x) = x^2 \cos x$ for $-6 \leq x \leq 6$

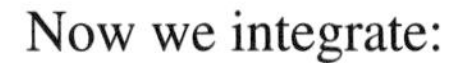
Now we integrate:

```
>> a = int(f,-6,6)

a =

68*sin(6)+24*cos(6)
```

To get a numerical result, we cast it with double:

```
>> double(a)

ans =

    4.0438
```

EXAMPLE 8-5
Calculate $\int_0^\infty e^{-x^2} \sin x \, dx$.

SOLUTION 8-5
We tell MATLAB that we want to evaluate the result at infinity by using inf as the upper limit:

```
>> a = int(exp(-x^2)*sin(x),0,inf)

a =

-1/2*i*pi^(1/2)*erf(1/2*i)*exp(-1/4)
```

Now we numerically evaluate the result:

```
>> double(a)

ans =

    0.4244
```

EXAMPLE 8-6
Find the volume of the solid of revolution obtained by rotating the curve e^{-x} about the x axis where $1 \le x \le 2$.

SOLUTION 8-6

The curve is shown in Figure 8-2. The volume of a solid generated by rotating a curve $f(x)$ about the x axis is given by:

$$\int_a^b \pi[f(x)]^2 dx$$

The integrand in this case is:

$$\pi(e^{-x})^2 = \pi e^{-2x}$$

The volume of the solid is then:

```
>> int(pi*exp(-2*x),1,2)

ans =

-1/2*pi*exp(-4)+1/2*pi*exp(-2)
```

Numerically we find this value is 0.1838.

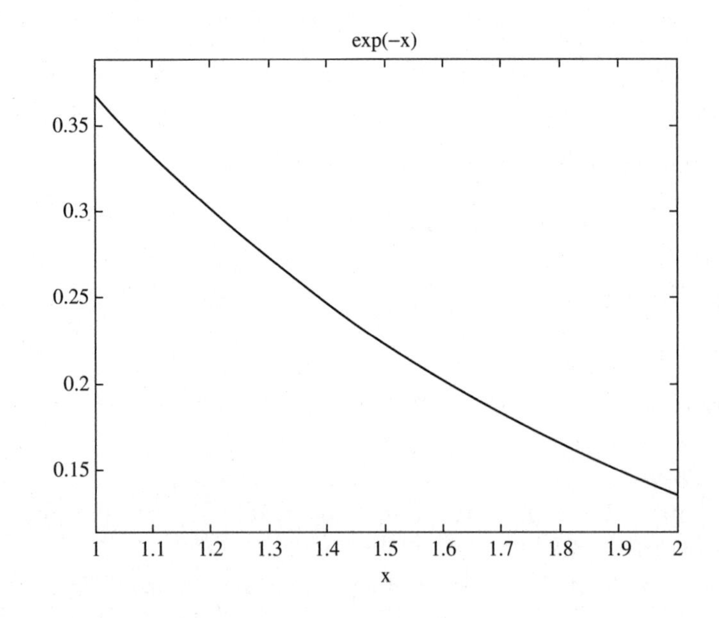

Figure 8-2 In Example 8-6 we find the volume of the solid of revolution generated by rotating the curve e^{-x} about the x axis

EXAMPLE 8-7

The sinc function, which is useful in signal processing for band-limited signals, is defined as:

$$f(x) = \frac{\sin(x)}{x}$$

Find the integral of the sinc function and sinc squared over the ranges $-20 \le x \le 20$, $-\infty < x < \infty$.

SOLUTION 8-7

Let's define both functions:

```
>> sinc = sin(x)/x;

>> sinc_squared = sinc^2;
```

The sinc-squared function has application for the description of the intensity of a light beam that has passed through a circular or square lens. First let's plot both functions, the sinc function first:

```
>> ezplot(sinc,[-20 20])
>> axis([-20 20 -0.5 1.2])
```

A plot of the sinc function is shown in Figure 8-3.

Now let's plot its square:

```
>> ezplot(sinc_squared,[-10 10])
>> axis([-10 10 -0.1 1.2])
```

A plot of the sinc-squared function is shown in Figure 8-4.

Now let's calculate the integrals for $-20 \le x \le 20$. For the sinc function we find:

```
>> a = int(sinc,-20,20)

a =

2*sinint(20)
```

We can cast this as a double to get a numerical result:

```
> double(a)

ans =

    3.0965
```

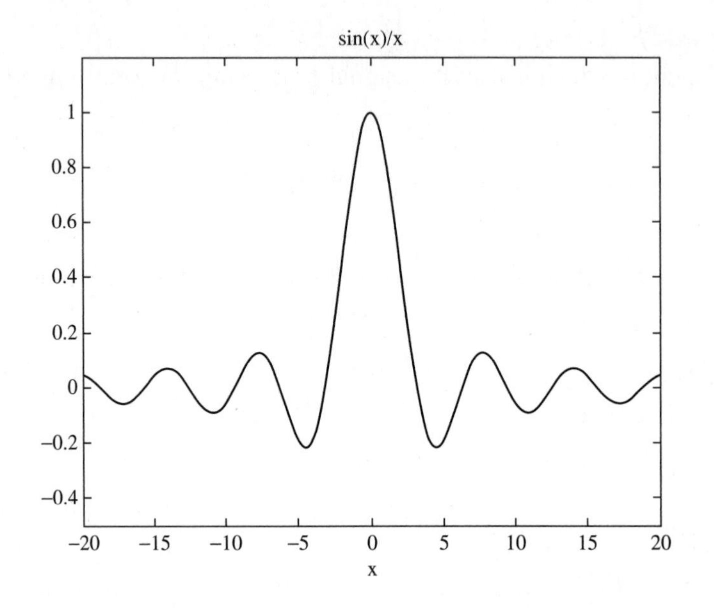

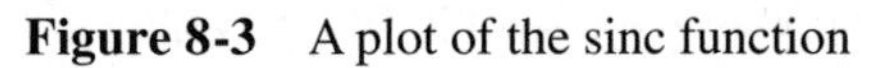

Figure 8-3 A plot of the sinc function

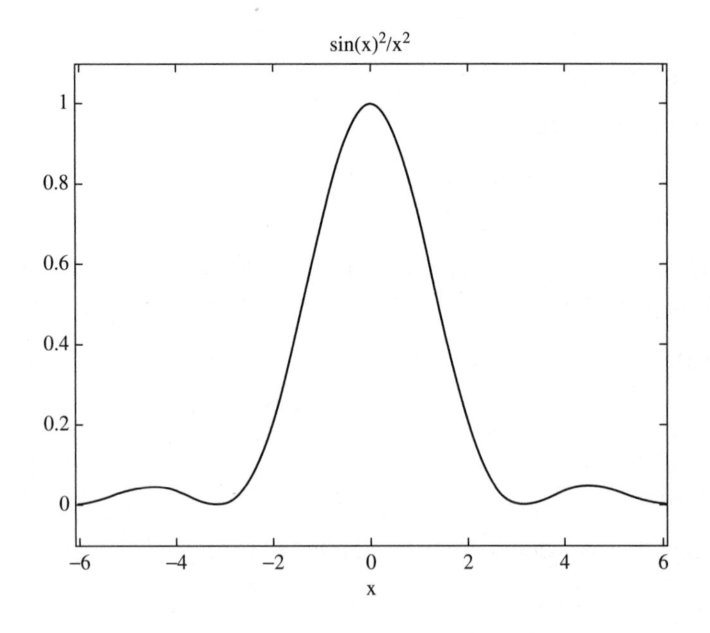

Figure 8-4 The sinc-squared function

Now we integrate sinc squared:

```
>> b = int(sinc_squared,-20,20)

b =

-1/20+1/20*cos(40)+2*sinint(40)
```

The numerical result is:

```
>> double(b)

ans =

    3.0906
```

Both results are very close. The integrals for $-\infty < x < \infty$ are:

```
>> a = int(sinc,-inf,inf)

a =

pi

>> b = int(sinc_squared,-inf,inf)

b =

pi
```

In fact what we find that as a gets bigger integrating over $-a \le x \le a$ both functions approach π. Sinc squared does so a little bit faster because the side lobes of the sinc function alternate between positive and negative and cancel out each other a little bit:

```
>> a = double(int(sinc,-50,50))

a =

    3.1032

>> b = double(int(sinc_squared,-50,50))

b =

    3.1217
```

Over the entire real line, this effect washes out and the functions both cover the same area.

Multidimensional Integration

We can compute multidimensional integrals in MATLAB by using nested int statements. Suppose that we wanted to compute the indefinite integral:

$$\int\int\int x y^2 z^5 dx\, dy\, dz$$

This can be done with:

```
>> syms x y z
>> int(int(int(x*y^2*z^5,x),y),z)

ans =

1/36*x^2*y^3*z^6
```

Definite integration proceeds analogously. We can calculate:

$$\int_1^2 \int_2^4 x^2 y\, dx\, dy$$

With the commands:

```
>> f = x^2*y;
>> int(int(f,x,2,4),y,1,2)

ans =

28
```

When computing multidimensional integrals in cylindrical and spherical coordinates, be sure to enter the correct area and volume elements—MATLAB will not do this automatically.

EXAMPLE 8-8

Find the volume of a cylinder of height h and radius a. What is the volume of a cylinder with radius $a = 3.5$ inches and height $h = 5$ inches?

SOLUTION 8-8

We will integrate using cylindrical coordinates with:

$$0 \le r \le a, \quad 0 \le \theta \le 2\pi, \quad 0 \le z \le h$$

The volume element in cylindrical coordinates is:

$$dV = r\,dr\,d\theta\,dz$$

So the volume of a cylinder of height h and radius a is:

$$V = \int_0^h \int_0^{2\pi} \int_0^a r\,dr\,d\theta\,dz$$

The commands to implement this in MATLAB are:

```
>> syms r theta z h a

>> V = int(int(int(r,r,0,a),theta,0,2*pi),z,0,h)

V =

a^2*pi*h
```

The volume of a cylinder with radius $a = 3.5$ inches and height $h = 5$ inches is:

```
>> subs(V,{a,h},{3.5,5})

ans =

   192.4226
```

The answer is expressed in cubic inches.

Numerical Integration

MATLAB can be used to perform trapezoidal integration by calling the trapz(x, y) function. Here x and y are two arrays, x containing the domain over which the integration takes place and y containing the function values at those points. Multiple functions can be integrated simultaneously (over the same domain x) by passing a multiple column argument y where each column contains the values of each function.

From calculus you are probably familiar with the trapezoidal method, which divides the region under a curve into a set of rectangles or panels. It then adds up the areas of the individual panels to get the integral.

EXAMPLE 8-9

Compute

$$\int_0^2 x^2 dx$$

using 10 and 20 evenly spaced panels, and compute the relative error in each case.

SOLUTION 8-9

First let's do the problem analytically to get the exact answer:

$$\int_0^2 x^2 dx = \frac{1}{3}x^3 \Big|_0^2 = \frac{8}{3} = 2.667$$

Now let's create a uniformly spaced line of ten panels:

```
>> x = linspace(0,2,10);
```

And define the function:

```
>> y = x.^2;
```

Integrating we find:

```
>> a = trapz(x,y)

a =

    2.6831
```

The relative error is:

```
>> 100*abs((8/3-a)/(8/3))

ans =

    0.6173
```

Now let's repeat the process with 20 uniformly spaced panels to see how we can reduce the error:

```
>> x = linspace(0,2,20);
>> y = x.^2;
>> a = trapz(x,y)

a =

    2.6704

>> 100*abs((8/3-a)/(8/3))

ans =

    0.1385
```

By doubling the number of panels, we have reduced the error by nearly a factor of 5.

EXAMPLE 8-10

Numerically estimate the following integrals of the Gaussian function:

$$\int_{-2}^{2} e^{-x^2} dx \text{ and } \int_{-\infty}^{\infty} e^{-x^2} dx$$

and compute the relative error. Use 200 uniformly spaced panels in the first case.

SOLUTION 8-10

Let's display our results using long format:

```
>> format long
```

In the first case we find:

```
>> x = linspace(-2,2,200);
>> gauss = exp(-x.^2);
>> in = trapz(x,gauss)

in =

   1.76415784847621
```

The actual result of this integral is

$$\int_{-2}^{2} e^{-x^2}\,dx = \sqrt{\pi}\,\mathrm{Erf}(2)$$

where Erf(2) is the *error function.* MATLAB evaluates this result to:

```
>> ac = sqrt(pi)*erf(2)

ac =

   1.76416278152484
```

We see that the results agree to four decimal places. The relative error, as a percentage, is:

```
>> d = ac - in;
>> err = abs(d/ac)*100

err =

    2.796254794950795e-004
```

For the second integral, the analytical result is:

$$\int_{-\infty}^{\infty} e^{-x^2}\,dx = \sqrt{\pi}$$

Let's plot the Gaussian to think about how to do the integral numerically. It is shown in Figure 8-5.

Generally, a Gaussian can be written in the form:

$$\varphi(x) = \exp\left(-\frac{(x-\mu)^2}{2\sigma^2}\right)$$

where μ is the mean and σ is the standard deviation. It turns out that about 99.73% of the area under a Gaussian curve falls within the range of three standard deviations. In our case we have $\exp(-x^2)$. The Gaussian is centered at the origin (so $\mu = 0$) and:

$$\frac{1}{2\sigma^2} = 1, \qquad \Rightarrow \sigma = \frac{1}{\sqrt{2}}$$

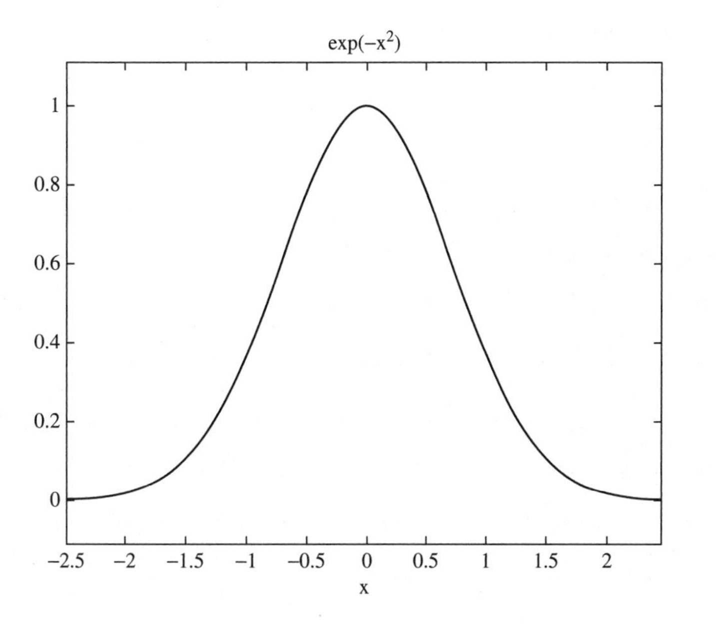

Figure 8-5 A Gaussian centered at the origin

This means that in our case, 99.73% of the area under the curve lies within the limits

$$x = \pm \frac{3}{\sqrt{2}} \approx \pm 2.213$$

So we can try doing this integral numerically by integrating over $-2.2 \le x \le 2.2$. Let's use 200 evenly spaced points over this range:

```
>> x = linspace(-2.2,2.2,200);
>> gauss = exp(-x.^2);
```

Integrating we obtain:

```
>> est = trapz(x,gauss)

est =

   1.76914920736586
```

The actual, analytic answer is:

```
>> ac = sqrt(pi)

ac =

    1.77245385090552
```

Our relative error is:

```
>> d = ac - est;
>> err = abs(d/ac)*100

err =

    0.18644454624112
```

Now 0.18% isn't a bad error considering the analytic result is found by integrating over the entire real line. Let's try extending our integration to $-3 \le x \le 3$:

```
>> x = linspace(-3,3,200);
>> gauss = exp(-x.^2);
>> est = trapz(x,gauss)

est =

    1.77241458438211
```

You can see that there has been an improvement in accuracy. Now the relative error is:

```
>> d = ac - est;
>> err = abs(d/ac)*100

err =

    0.00221537634854
```

At 0.002 % this estimate is quite a bit more accurate.

EXAMPLE 8-11

The velocity of a rocket sled on a track is measured once a second for 10 seconds. The velocity in ft/s is found to be:

t	1	2	3	4	5	6	7	8	9	10
$v(t)$	65	95	110	150	170	210	240	260	265	272

Find the total distance covered by the sled.

SOLUTION 8-11

Velocity is related to position by:

$$v = \frac{dx}{dt}$$

Hence we can find the position from the velocity by integrating:

$$x(t) = \int_a^b v(t)\,dt$$

To do this numerically, first let's put the data into two arrays:

```
>> t = [1:1:10];
>> v = [65 95 110 150 170 210 240 260 265 272]
```

We can calculate the value of each element from:

$$x(t_{k+1}) = \int_{t_k}^{t_{k+1}} v(t)\,dt + x(t_k)$$

We initialize the first element to zero:

```
>> x(1) = 0;
```

Now we do the integration using a for loop:

```
>> for k = [1:9]
x(k+1) = trapz(t(k:k+1),v(k:k+1))+x(k);
end
```

The result is:

Time	Position (in 1,000 feet)
1	0
2	0.08000000000000
3	0.18250000000000
4	0.31250000000000
5	0.47250000000000
6	0.66250000000000
7	0.88750000000000
8	1.13750000000000
9	1.40000000000000
10	1.66850000000000

So the sled has traveled 1668.50 feet. This technique gives us the position at each second. If we just wanted the position at the end, the trapz function can give us that in one call:

```
>> trapz(t,v)

ans =

    1.668500000000000e+003
```

Quadrature Integration

MATLAB has two commands called *quad* and *quad*l that can be used to perform *quadrature* integration. This type of method is based on the idea that you can approximate the area under a curve better using quadratic functions instead of rectangles (even more accuracy can be obtained by using higher order polynomials). Simpson's rule does this by dividing the range of integration up into an even number of sections, approximating the area under each pair of adjacent sections by a different quadratic function. The quad function uses an adaptive Simpson's rule approaches to do numerical integration. To use quad, you pass the function to be integrated along with the limits of integration.

The quadl function uses *Lobatto* integration. This is a more sophisticated type of adaptive quadrature integration, see the MATLAB help for more information.

The downside of the quad and quadl functions is that they cannot integrate a set of points.

EXAMPLE 8-12

Use quadrature integration to compute

$$\int_0^{1/8} e^{-2x} dx$$

SOLUTION 8-12

The analytical answer to four decimal places is:

$$\frac{1}{2} - \frac{1}{2e^{1/4}} \approx 0.1106$$

The quad function returns:

```
> quad('exp(-2*x)',0,1/8)

ans =

    0.1106
```

In fact if we carry the answer to 14 decimal places we still find excellent agreement:

```
>> format long

>> ac = 1/2 - 1/(2*exp(1/4))

ac =

   0.11059960846430

>> quad('exp(-2*x)',0,1/8)

ans =

   0.11059960846436
```

Using the trapz function we find a good answer but it differs in the seventh decimal place:

```
> x = linspace(0,1/8,100);
>> y = exp(-2*x);
>> r = trapz(x,y)

r =

   0.11059966723785
```

Quiz

1. Compute $\int x e^{-ax}\, dx$.
2. Compute $\int \frac{dx}{\sqrt{x^2+bx+c}}$.
3. Compute $\int x \tan x\, dx$.

4. Find $\int \sin^2(yx)dy$.
5. Compute $\int_0^\infty e^{-3x} \cos x \, dx$.
6. Find the volume of the solid of revolution obtained by rotating the region between the curves $\sqrt{x}$ and x^2 for $0 \le x \le 1$.
7. Use MATLAB to generate the formulas for the surface area and volume of a sphere of radius a. What is the volume of a sphere with $a = 2$ m?
8. Numerically calculate $\int_0^\pi \cos(\pi x)dx$ using trapezoidal integration. Use 50 evenly spaced panels and determine the relative error.
9. Numerically integrate the Gaussian over $-2.2 \le x \le 2.2$, using 1000 evenly spaced panels. Does this reduce the relative error by a significant amount?
10. Numerically compute:

$$\int_0^{10} J_1(x)\,dx$$

where $J_1(x)$ is the Bessel function of the first kind using trapezoidal and quadrature methods.

CHAPTER 9

Transforms

Transforms, such as the Laplace, *z*, and Fourier transforms, are used throughout science and engineering. Besides simplifying the analysis, transforms allow us to see data in a new light. For example, the Fourier transform allows you to view a signal that was represented as a function of time as one that is a function of frequency. In this chapter we will introduce the reader to the basics of using MATLAB to work with transforms. In this chapter we will introduce the laplace, fourier, and fft commands.

The Laplace Transform

The *Laplace transform* of a function of time $f(t)$ is given by the following integral:

$$\ell\{f(t)\} = \int_0^\infty f(t)\, e^{-st}\, dt$$

We often denote the Laplace transform of $f(t)$ by $F(s)$. The advantage of the Laplace transform is that it turns differential equations into algebraic ones. In principle, an algebraic equation should be much easier to solve, but this doesn't always work out in practice. But with MATLAB at our disposal the situation is greatly simplified.

To compute a Laplace transform in MATLAB, we make a call to laplace($f(t)$). This is done using symbolic computation. Let's use this function to build up a list of Laplace transforms that you can find in any differential equations or electrical engineering book.

We will start with the easiest example, computing the Laplace transform of the constant function $f(t) = a$. First we define our symbolic variables:

```
>> syms s t
```

First watch what happens if we try to compute the Laplace transform of the number 1:

```
>> laplace(1)
??? Function 'laplace' is not defined for values of class 'double'.
```

Hence we need to define a symbolic constant first:

```
>> syms a
```

Now we can find the Laplace transform of a constant:

```
>> laplace(a)

ans =

1/s^2
```

Here are the Laplace transforms of some higher powers of t:

```
>> laplace(t^2)

ans =

2/s^3

>> laplace(t^7)

ans =

5040/s^8
```

```
>> laplace(t^5)

ans =

120/s^6
```

From this we deduce the well-known formula:

$$\ell\{t^n\} = \frac{n!}{s^{n+1}}$$

Now let's look at the Laplace transform of some common functions seen in science and engineering. First consider the decaying exponential:

```
>> laplace(exp(-b*t))

ans =

1/(s+b)
```

The Laplace transforms of the sin and cos functions are:

```
>> laplace(cos(w*t))

ans =

s/(s^2+w^2)

>> laplace(sin(w*t))

ans =

w/(s^2+w^2)
```

We can also calculate the Laplace transform of the hyperbolic cosine function:

```
>> laplace(cosh(b*t))

ans =

s/(s^2-b^2)
```

The Laplace transform is linear. That is:

$$\ell\{af(t) + bg(t)\} = a\ell\{f(t)\} + b\ell\{g(t)\}$$

Let's verify this with MATLAB. Let

```
>> f = 5 + exp(-2*t)
```

We find that the Laplace transform is:

```
>> laplace(f)

ans =

5/s+1/(s+2)
```

The Inverse Laplace Transform

Computing inverse Laplace transforms is something sure to generate a lot of headaches among readers who are taking say a circuits analysis class. This process is often very tedious and filled with lots of difficult algebra. But with MATLAB our problems are solved. We can compute the inverse Laplace transform by typing *ilaplace*, saving us from having to do partial fraction decompositions and other nasty tricks. First let's get familiar with computing inverse Laplace transforms by looking at some simple examples. First consider a simple power function:

```
>> ilaplace(1/s^3)

ans =

1/2*t^2
```

Or with this one we get an exponential:

```
>> ilaplace(2/(w+s))

ans =

2*exp(-w*t)
```

And finally an example involving a trig function:

```
>> ilaplace(s/(s^2+4))

ans =

cos(2*t)
```

Of course most problems you come across in applications don't result in inverse Laplace transforms where you can just spot the answer like in these examples. This is where MATLAB is really useful. For instance, what is the inverse Laplace transform of:

$$F(s) = \frac{5-3s}{2+5s}$$

Let's enter it into MATLAB:

```
>> F = (5-3*s)/(2+5*s);
```

Computing the inverse Laplace transform, we find a Dirac delta function (or *unit impulse* function for engineers):

```
>> ilaplace(F)

ans =

-3/5*dirac(t)+31/25*exp(-2/5*t)
```

Here is a more complicated example. What do you think the inverse Laplace transform of:

$$F(s) = \frac{-s^2 - 9s + 4}{s^2 + s + 2}$$

happens to be? Well, you can start involving yourself in tedious algebra to get it into a form that is immediately recognizable or you can enter it in MATLAB:

```
>> F = (-s^2 - 9*s + 4)/(s^2 + s + 2);
```

Here is the result:

```
>> ilaplace(F)

ans =

-dirac(t) -8*exp(-1/2*t)*cos(1/2*7^(1/2)*t)+20/7*7^(1/2)*
exp(-1/2*t)*sin(1/2*7^(1/2)*t)
```

Kind of messy—but at least we have the answer! Let's try some examples you are likely to come across involving partial fractions. For instance:

```
>> F = 1/(s*(s+1)*(s+2));
>> ilaplace(F)

ans =

-exp(-t)+1/2*exp(-2*t)+1/2
```

EXAMPLE 9-1

What function of time corresponds to:

$$F(s) = \frac{1}{(s+7)^2}$$

Plot the function.

SOLUTION 9-1

The inverse laplace command gives:

```
>> f = ilaplace(1/(s+7)^2)

f =

t*exp(-7*t)
```

Remember, we are using symbolic computing here. So we can plot it using ezplot:

```
>> ezplot(f)
```

The result is shown in Figure 9-1. In an application, we are probably only interested in the behavior of the function for $t \geq 0$.

So let's try looking at it for positive values. First let's try looking at the function for the first 5 seconds. We do this by typing:

```
>> ezplot(f, [0,5])
```

This plot is shown in Figure 9-2. It seems the function quickly goes to zero and all the action is prior to 1 second, so this plot is also unsatisfying. So we try once more, looking only at $0 \leq t \leq 1$:

```
>> ezplot(f, [0,1])
```

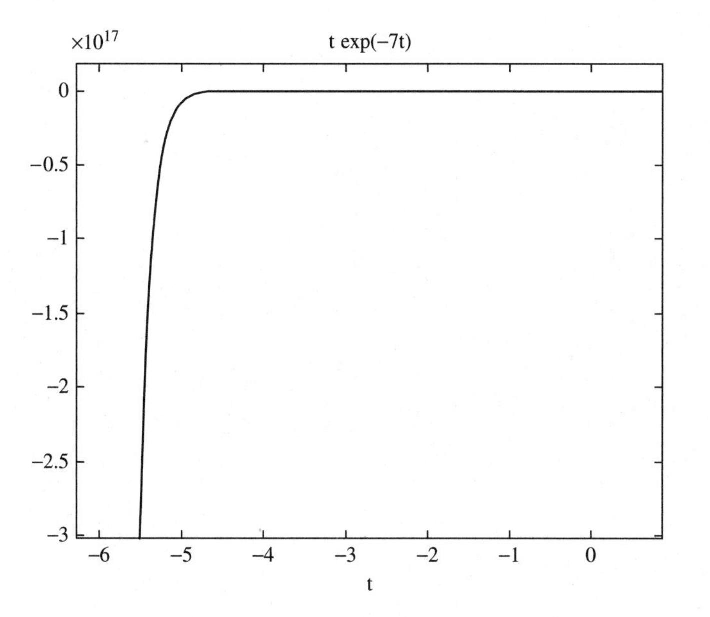

Figure 9-1 A plot of te^{-7t}

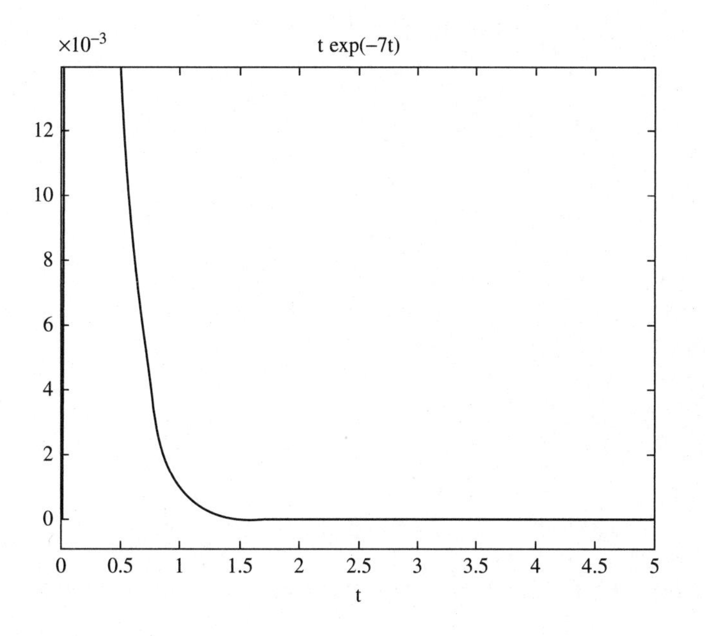

Figure 9-2 Looking at te^{-7t} for $0 \le t \le 5$ seconds

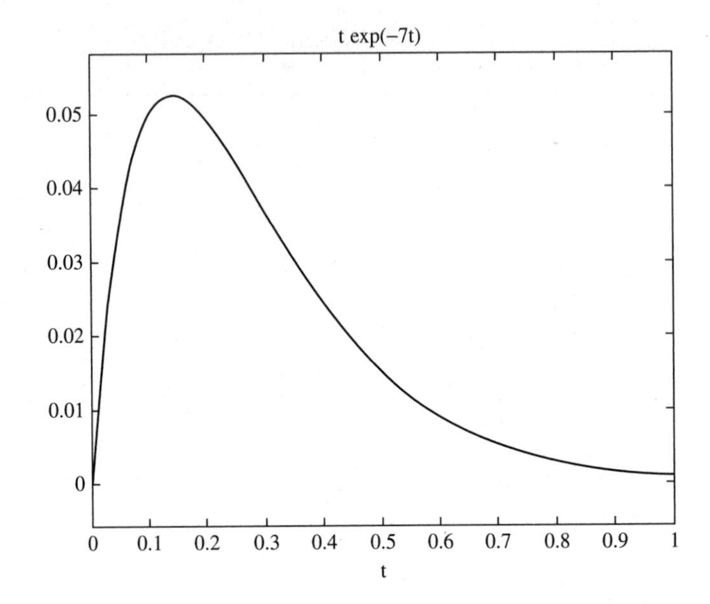

Figure 9-3 A plot of the function with a more suitable range of time

The result this time, shown in Figure 9-3, gives a nice view of the function.

EXAMPLE 9-2

Find and plot the function of time that corresponds to:

$$\frac{2s+3}{(s+1)^2(s+3)^2}$$

SOLUTION 9-2

Calling ilaplace we find:

```
>> f = ilaplace((2*s+3)/((s+1)^2*(s+3)^2))

f =

exp(-2*t)*(-1/2*t*cosh(t)+(1/2+t)*sinh(t))
```

In this case, we find that plotting over a range of 7 seconds produces a nice plot:

```
>> ezplot(f,[0,7])
```

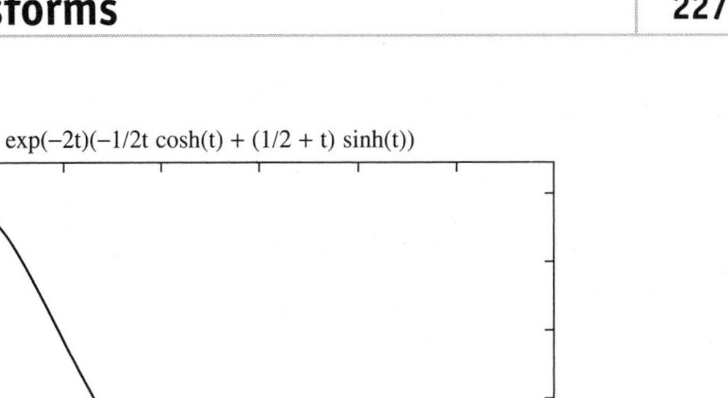
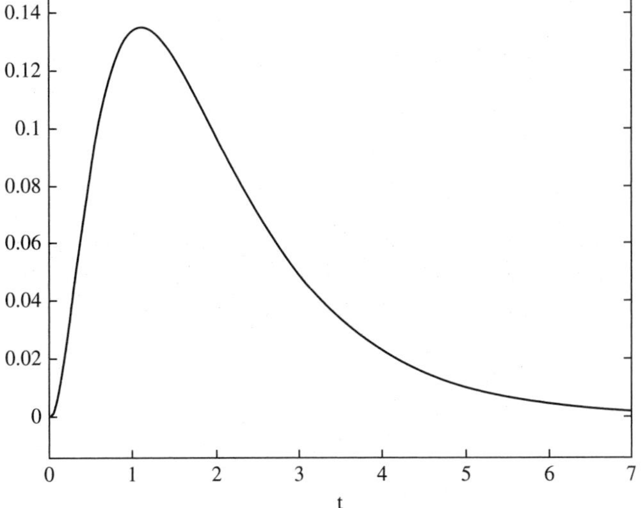

Figure 9-4 A plot of the function used in Example 9-2

The important idea is to generate a plot that brings out the important features of the function for $t < 0$. The plot is shown in Figure 9-4. Notice that the function has the same shape that we saw in the previous example—but that the response has been stretched out over a much longer time interval.

Solving Differential Equations

The Laplace transform simplifies differential equations, turning them into algebraic ones. The Laplace transform of the first derivative of a function is:

$$\ell\left\{\frac{df}{dt}\right\} = sF(s) - f(0)$$

And the Laplace transform of the second derivative of a function is:

$$\ell\left\{\frac{d^2 f}{dt^2}\right\} = s^2 F(s) - sf(0) - f(0)$$

Unfortunately MATLAB doesn't work quite the way I would like. Consider the fact that you enter a derivative for the DSOLVE command by placing a D in front of the variable. For example we can find the solution to a first order ODE:

```
>> dsolve('Dx = a*x')

ans =

C1*exp(a*t)
```

It would be nice if you could have Laplace generate the results stated above for derivatives, but it can't. For example, if we type:

```
>> laplace(Dx -a*x)
```

We get an error message:

```
??? Undefined function or variable 'Dx'.
```

Don't try entering it as a character string:

```
>> laplace('Dx -a*x')
??? Function 'laplace' is not defined for values of class 'char'.
```

So we can't use Laplace to work directly on derivatives. To solve a differential equation with MATLAB using Laplace transform methods, you will have to compute the Laplace transform yourself and enter the result in MATLAB, and then use ilaplace to come up with a solution. We illustrate with an example.

EXAMPLE 9-3

Consider two mass-spring systems where $m = 1$ kg that are described by the following differential equation:

$$m\ddot{x}(t) + 1.2\dot{x}(t) + x = y(t) + \alpha\dot{y}(t)$$

Find a solution of the equation $X(s)$in the s domain and invert to find $x(t)$. Plot for the three values of the constant $\alpha = 0, 2, 5$. Let $x(0) = \dot{x}(0) = y(0)$. Suppose that $y(t)$is given by the unit step or *Heaviside* function.

SOLUTION 9-3

Using the rule for the Laplace transform of the first and second derivatives, and setting $m = 1$, we arrive at the following equation in s:

$$s^2X(s) + 1.2sX(s) + X(s) = Y(s) + sY(s)$$

Some rearrangement gives:

$$(s^2 + 1.2s + 1)X(s) = (1 + \alpha s)Y(s)$$

$$\Rightarrow X(s) = \frac{1 + \alpha s}{s^2 + 1.2s + 1} Y(s)$$

In the problem statement we are told that $y(t)$ is the unit step or Heaviside function. For readers not familiar with this function, it is defined as:

$$y(t) = \begin{cases} 0 & t < 0 \\ 1 & t \geq 1 \end{cases}$$

We can plot it in MATLAB with the following command:

```
>> ezplot(Heaviside(t),[-2,2])
```

The result is shown in Figure 9-5. Essentially we have a system with a constant forcing function turned on at $t = 0$.

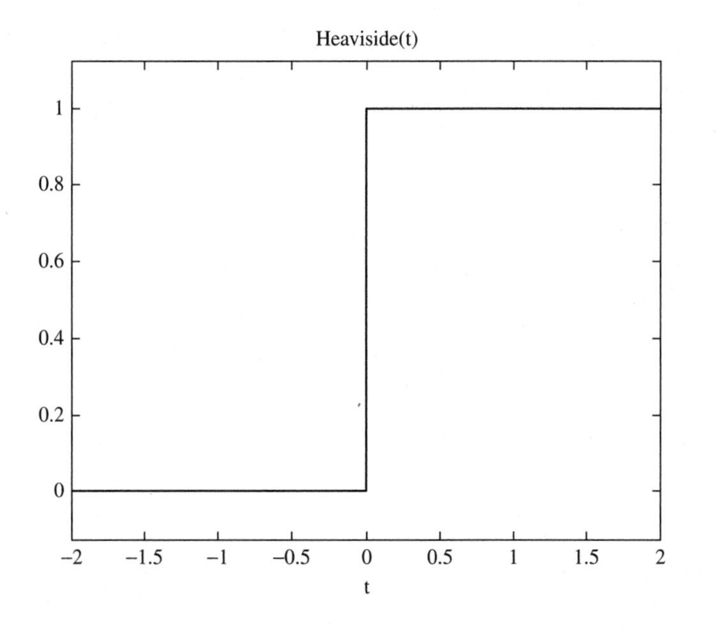

Figure 9-5 A plot of the Heaviside or unit step function

To solve the problem, we need to know the Laplace transform of the Heaviside function. It is:

```
>> laplace(heaviside(t))

ans =

1/s
```

Hence the function $x(t)$ can be found by inverting:

$$X(s) = \frac{1+\alpha s}{s^2 + 1.2s + 1}\left(\frac{1}{s}\right)$$

Let's label our three cases by $\alpha = a, b, c$ where $a = 0, b = 2, c = 5$:

```
>> a = 0; b = 2; c = 5;
```

Now we enter our function in the s domain, with the three cases. First, to simplify typing, let's enter the denominator that is the same in all three cases:

```
>> d = s^2 + (1.2)*s + 1;
```

We can then define three functions of s for each of the constants:

```
>> Xa = ((1+ a*s)/d)*(1/s);
>> Xb = ((1+ b*s)/d)*(1/s);
>> Xc = ((1+ c*s)/d)*(1/s);
```

Now we invert:

```
>> xa = ilaplace(Xa)

xa =

-exp(-3/5*t)*cos(4/5*t)-3/4*exp(-3/5*t)*sin(4/5*t)+1


>> xb = ilaplace(Xb)

xb =

1-exp(-3/5*t)*cos(4/5*t)+7/4*exp(-3/5*t)*sin(4/5*t)

>> xc = ilaplace(Xc)

xc =

1-exp(-3/5*t)*cos(4/5*t)+11/2*exp(-3/5*t)*sin(4/5*t)
```

The cases $\alpha = 2$, $\alpha = 5$ are similar, so let's just plot the $\alpha = 2$ case side by side with $\alpha = 0$. We can do this using the subplot command. First let's tell it that we will have 1 row with 2 panes, and that we are putting our first plot in pane 1:

```
>> subplot(1,2,1)
```

We plot the first function for $0 \le t \le 10$:

```
>> ezplot(xa, [0 10])
```

Now we tell MATLAB where to put the next plot:

```
>> subplot(1,2,2)
```

And we plot the second function for $0 \le t \le 10$:

```
>> ezplot(xb, [0 10])
```

The results are shown in Figure 9-6.

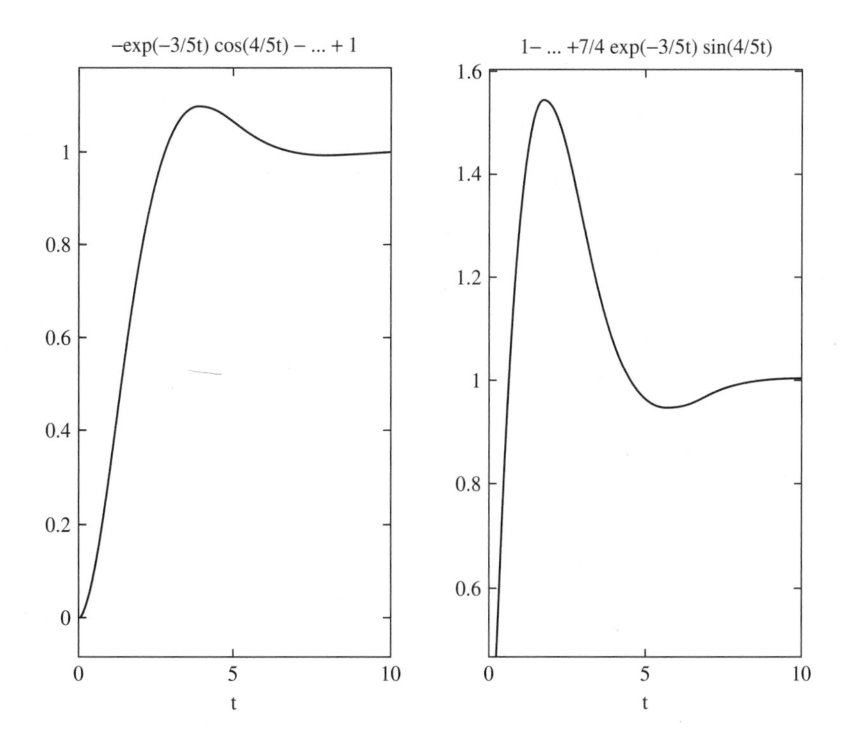

Figure 9-6 Plotting solutions of the differential equation in Example 9-3

Computing Fourier Transforms

The *Fourier transform* of a function *f*(*t*) is defined as:

$$F(\omega) = \int_{-\infty}^{\infty} f(t)\, e^{-i\omega t} dt$$

We can calculate the Fourier transform of a function in MATLAB by typing the command fourier. A Fourier transform allows you to convert a function of time or space into a function of frequency. For example, we can verify that the Fourier transform of the sin function is given by two Dirac deltas:

```
>> fourier(sin(x))

ans =

i*pi*(-dirac(w-1)+dirac(w+1))
```

Here we find the Fourier transform of a Gaussian. First let's define the function and plot it in the "spatial" domain:

```
>> f = exp(-2*x^2);

>> ezplot(f,[-2,2])
```

The plot is shown in Figure 9-7.
We compute the Fourier transform:

```
>> FT = fourier(f)

FT =

1/2*2^(1/2)*pi^(1/2)*exp(-1/8*w^2)
```

We have found the famous result that the Fourier transform of a Gaussian is another Gaussian—albeit with some scaling parameters. Let's look at the plot that shows some nice features of Fourier transforms. The function is wider in frequency and has a higher peak, as can be seen when comparing Figure 9-8 with Figure 9-7.

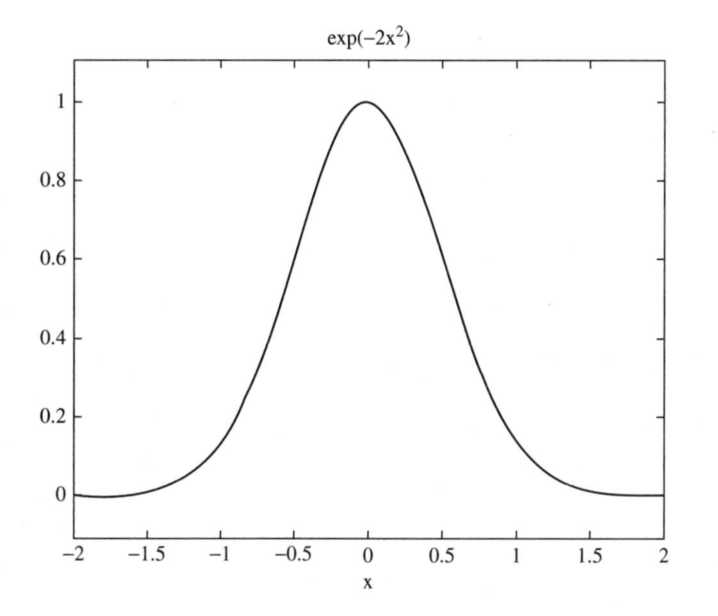

Figure 9-7 A plot of our function in the spatial domain

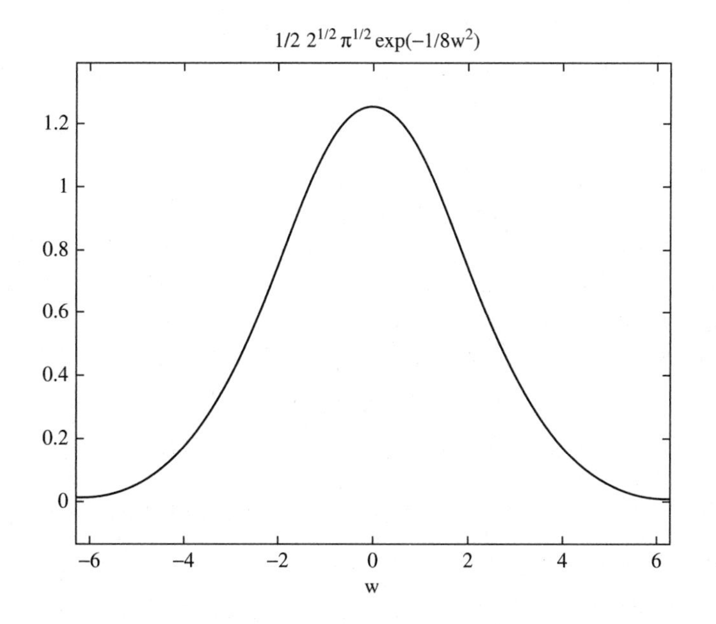

Figure 9-8 The Fourier transform of a Gaussian is another Gaussian with a different height and width

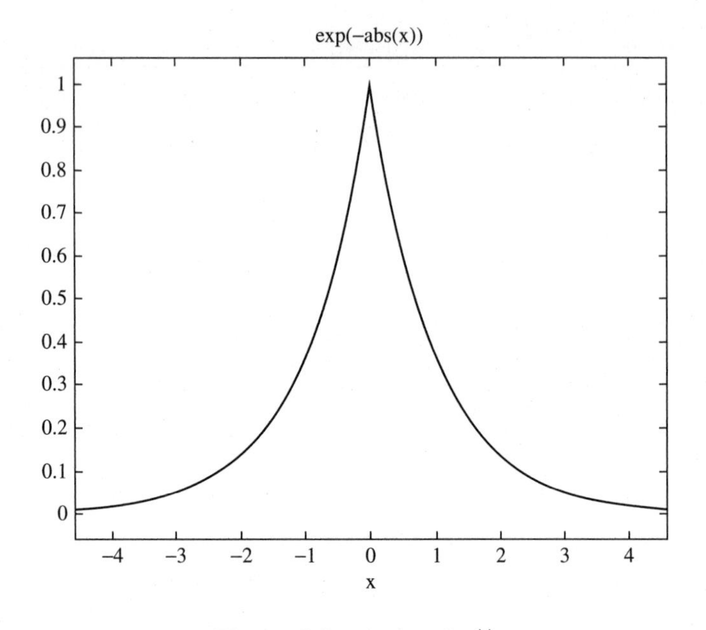

Figure 9-9 A plot of $e^{-|x|}$

Here is another nice example. Consider the function:

$$f(x) = e^{-|x|} = \begin{cases} e^{x} & x < 0 \\ e^{-x} & x > 0 \end{cases}$$

We can define it in MATLAB by typing:

```
>> f = exp(-abs(x));
```

Let's plot it. The result is shown in Figure 9-9.
Now we compute the Fourier transform:

```
>> FT = fourier(f)

FT =

2/(w^2+1)
```

A plot of this function of frequency is shown in Figure 9-10.

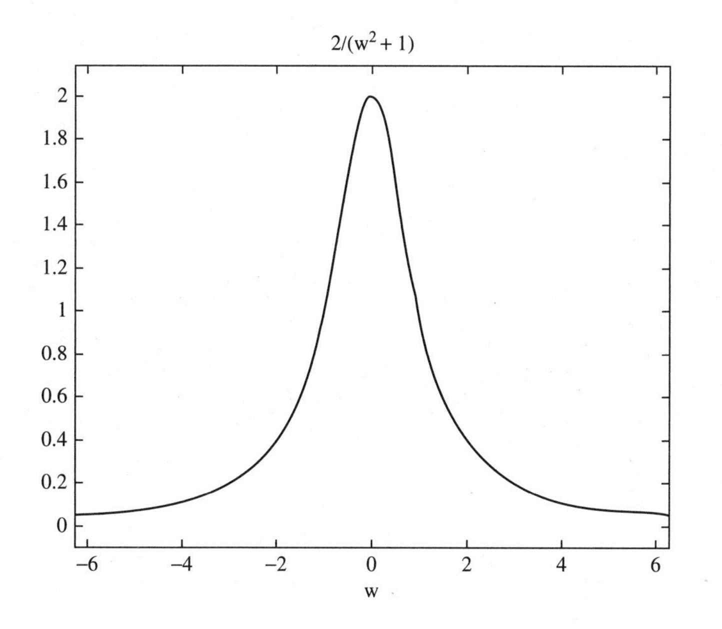

Figure 9-10 A plot of the Fourier transform of $e^{-|x|}$

Inverse Fourier Transforms

To compute the inverse Fourier transform of a function, you can use the *ifourier* command. For example, we can see the duality relationship of the Fourier transform by typing:

```
>> f = ifourier(-2*exp(-abs(w)))
```

The result is:

```
f =

-2/(x^2+1)/pi
```

This function is shown in Figure 9-11.

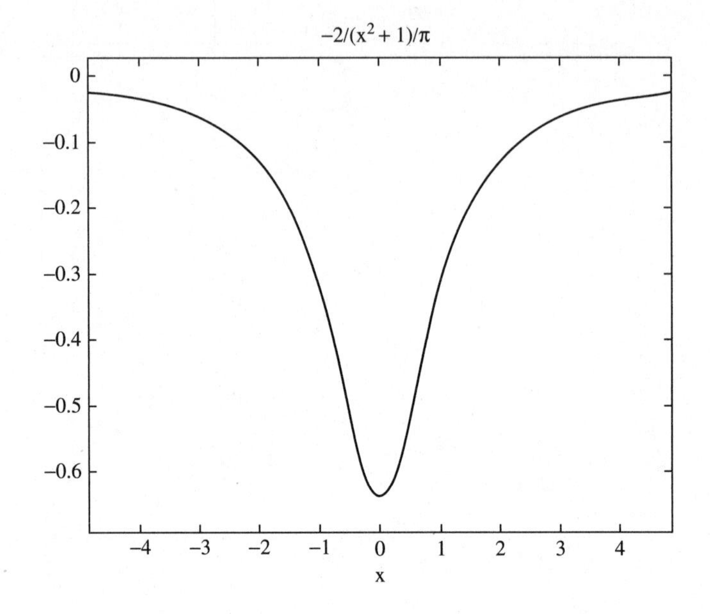

Figure 9-11 Plot of the function obtained by computing the inverse Fourier transform of $f(\omega) = -2e^{-|\omega|}$

Fast Fourier Transforms

The MATLAB function *fft* can be used to compute numerical fast fourier transforms of vectors of numbers.

EXAMPLE 9-4

Suppose that a signal $x(t) = 3\cos(\pi t) + 2\cos(3\pi t) + \cos(6\pi t)$. Develop a model to describe this signal corrupted by noise, and compute the FFT to examine the frequency content of the signal. Consider a time interval of 10 seconds.

SOLUTION 9-4

First let's define our interval:

```
>> t = 0:0.01:10;
```

Now let's define the signal:

```
>> x = 3*cos(pi*t) + 2*cos(3*pi*t) + cos(6*pi*t);
```

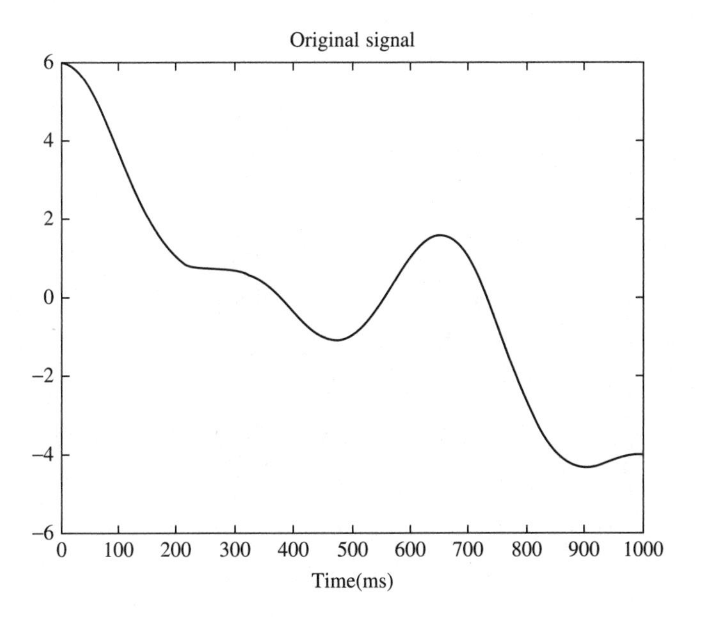

Figure 9-12 The original signal

To model noise, we will generate some random numbers and add them into the signal. This can be done with a call to randn:

```
>> x_noisy = x + randn(size(t));
```

Let's plot the two functions to see the effect of the noise. In Figure 9-12, we show the original uncorrupted signal over the first 1000 milliseconds. The command used to plot is:

```
>> plot(1000*t(1:100),x(1:100)), xlabel('time (ms)'),
title('Original Signal')
```

Now let's plot the noisy signal over the same interval. The command is:

```
>> plot(1000*t(1:100),x_noisy(1:100)),xlabel('time(ms)'),
title('Noisy Signal')
```

The result is shown in Figure 9-13.

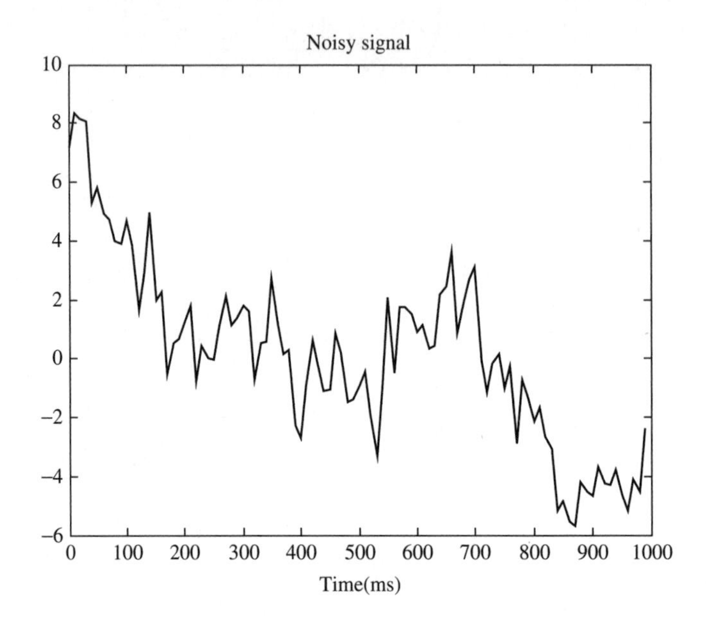

Figure 9-13 The noisy version of our signal

The noisy signal looks hopeless. We can glean some useful information about the original signal by computing the Fourier transform. We can compute an n-point fast Fourier transform of a function f by typing *fft(f, n)*. Following an example given in the MATLAB help, we compute the 512-point Fourier transform of our function:

```
>> FT = fft(x_noisy,512);
```

The power in the signal can be found by taking the product of the Fourier transform with its complex conjugate, and dividing by the total number of points:

```
>> P = FT.*conj(FT)/512;
```

When we plot the function, the frequency content of the signal is apparent. In Figure 9-14 we see three peaks, corresponding to the three frequencies in $x(t) = 3\cos(\pi t) + 2\cos(3\pi t) + \cos(6\pi t)$. Notice how the amplitudes in the original signal are reflected in the power spectrum of Figure 9-14.

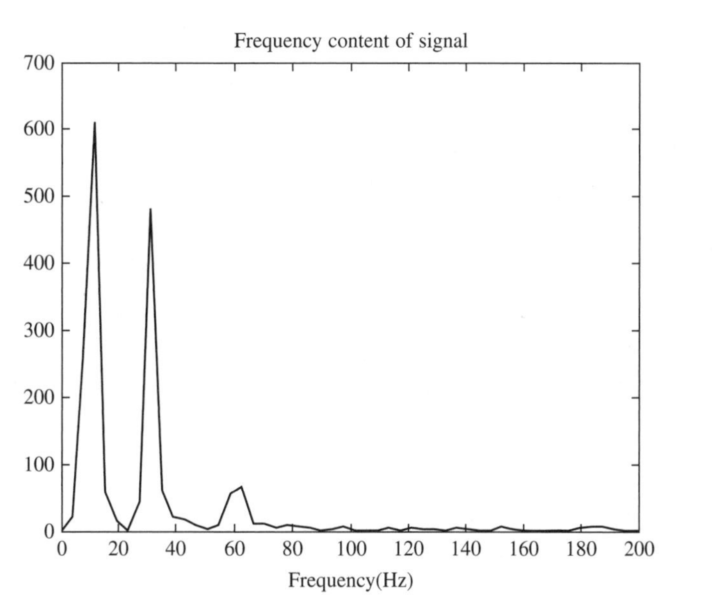

Figure 9-14 By taking the discrete Fourier transform of the noisy signal, we are able to extract some frequency information about it. The three peaks and their relative strengths are reflective of the frequency content of the signal $x(t) = 3\cos(\pi t) + 2\cos(3\pi t) + \cos(6\pi t)$

Quiz

1. Find the Laplace transform of $g(t) = te^{-3t}$.
2. Find the Laplace transform of $f(t) = 8\sin 5t - e^{-t}\cos 2t$.
3. What is the inverse Laplace transform of $\frac{1}{s} - \frac{2s^2+3}{s^2+9}$.
4. Find the inverse Laplace transform of $\frac{s}{s^2-49} - \frac{3}{s^2-9}$.
5. Solve the differential equation:

$$\frac{dy}{dt} + y = 2U(t)$$

where $U(t)$ is the Heaviside function. Let all initial conditions be zero. What is the forced response?

6. Find the Fourier transform of x^2.
7. Find the Fourier transform of $x \cos x$.
8. Find the inverse Fourier transform of $\frac{1}{1+i\omega}$.
9. What is the fast Fourier transform of $a = [2,4,-1,2]$?
10. Let $x(t) = \sin(\pi t) + 2\sin(4\pi t)$. Add some noise to the signal using x + randn(size(t)). What is the highest power content of the signal and at what frequency?

CHAPTER 10

Curve Fitting

MATLAB can be used to find an appropriate function that best fits a set of data. In this chapter we will examine some simple techniques that can be used for this purpose.

Fitting to a Linear Function

The simplest case we can imagine is a data set which is best described by a linear function. That is if our data is of the form $y = f(x)$ we are considering the case where $f(x)$ is such that:

$$y = mx + b$$

To find the values of m and b, we can apply a MATLAB function called polyfit(x,y,n) where n is the degree of the polynomial we want MATLAB to find. For an equation of the form $y = mx + b$, we set n equal to unity and the call would be polyfit(*x, y,* 1). The polyfit function works by using least squares. Let's see how to apply it with some simple examples.

EXAMPLE 10-1

Among a set of golfers, a relationship between handicap and average score is noted as follows:

Handicap	6	8	10	12	14	16	18	20	22	24
Average	3.94	3.8	4.1	3.87	4.45	4.33	4.12	4.43	4.6	4.5

Find a curve that fits the data and gives an estimate of the goodness of the fit.

SOLUTION 10-1

An understanding of golf isn't necessary for this example, all you need to know is we are assuming a linear relationship between handicap and average, and we want to derive an equation to describe it. Once you have the equation $y = mx + b$, you can predict values of y for values of x you don't yet have. First let's enter our data in two arrays:

```
>> handicap = [6:2:24]

handicap =

     6     8    10    12    14    16    18    20    22    24

>> Ave = [3.94, 3.8, 4.1, 3.87, 4.45, 4.33, 4.12, 4.43, 4.6, 4.5];
```

Next we call polyfit to have MATLAB generate the coefficients for a polynomial to fit the data. To have MATLAB generate a first order polynomial of the form $y = mx + b$, first we need to determine what is x (the independent variable) and what is y (the dependent variable). In this case the independent variable which plays the role of x is the Handicap, and the dependent variable which plays the role of y is the Average. Since we want to generate a first degree polynomial, we call polyfit in the following way:

```
>> p = polyfit(handicap,Ave,1);
```

Next we need to extract the coefficients that MATLAB finds. In general, polyfit will generate the coefficients for the polynomial in the following way:

$$p(x) = p_1 x^n + p_2 x^{n-1} + \cdots + p_{n-1} x^2 + p_n x + p_{n+1}$$

If we make a call p = polyfit(x, y, n) the jth coefficient is referenced by writing $p(j)$. In our case, polyfit returns coefficients for an equation of the form $p(1)*x + p(2)$. Therefore we can extract the coefficients this way:

```
> m = p(1)

m =

    0.0392

>> b = p(2)

b =

    3.6267
```

Now let's generate a function to draw the line $y = mx + b$. First we need to represent the x axis:

```
x = [6:0.1:24];
```

Now we create the function to draw the line:

```
>> y = m*x + b;
```

Let's plot the line along with the actual data, which will be shown as individual points. We can represent the actual data with circles using the following commands:

```
>> subplot(2,1,2);

>> plot(handicap,Ave,'o',x,y),xlabel('Handicap'),ylabel('Average')
```

The result is shown in Figure 10-1. The fit is not perfect, many of the data points are scattered quite a bit off the line we generated. Let's see what the fit generated in this case predicts for the specific handicap values for which we have real data. This can be done by executing the following command:

```
>> w = m*handicap + b

w =

    3.8616    3.9399    4.0182    4.0965    4.1748    4.2532
4.3315    4.4098    4.4881    4.5664
```

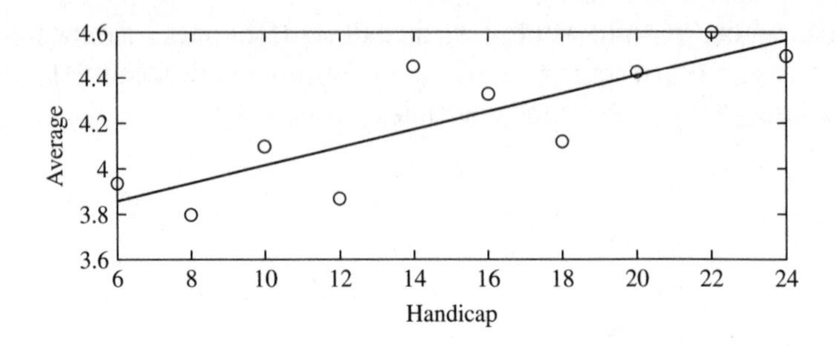

Figure 10-1 A plot of the data along with the least squares fit generated in Example 10-1

Looking back at the original table of data, we can see that this is a pretty good approximation. Since this data is generated using the smooth line $y = mx + b$ generated to fit the data, the data stored in w is often called *smoothed* data.

In many cases observational data is plagued with errors, and the equation $y = mx + b$ may be taken to be a better representation of the actual relationship between the dependent and independent variables than the collected data is.

Now let's take a look at some ways to characterize how good the fit really is. This can be done by getting an estimate of the error of the fit. The first item we can use to look at how good the fit is are the residuals. Suppose that we are fitting a function $f(x)$ to a set of collected data t_i. The sum of the squares of the residuals is given by:

$$A = \sum_{i=1}^{N} [f(x_i) - t_i]^2$$

Now let the collected data have a mean or average value given by $\bar{t}$. The sum of the squares of the deviation of the collected data from the mean is:

$$S = \sum_{i=1}^{N} (t_i - \bar{t})^2$$

The r-squared value is then:

$$r^2 = 1 - \frac{A}{S}$$

If $r^2 = 1$, then the function would be a perfect fit to the data. Hence the closer r^2 is to 1, the better the fit. In MATLAB, we can implement these formulas pretty easily. First let's get the mean of our collected data. There are ten data points, hence:

```
>> N = 10;
>> MEAN = sum(Ave)/N

MEAN =

    4.2140
```

We can also calculate the mean of a set of data stored in an array using the built-in function:

```
>> mean(Ave)

ans =

    4.2140
```

Then:

```
>> S = sum((Ave-MEAN).^2)

S =

    0.7332
```

(Don't be confused by the use of Ave, which is the array containing the average golf scores). Now we compute A. To do this, we need the values of our fit at the relevant data points—we did this by creating the array *w*. So:

```
>> A = sum((w-Ave).^2)

A =

    0.2274
```

Now we compute r^2:

```
>> r2 = 1 - A/S

r2 =

    0.6899
```

The r-squared value is not so close to 1, so this isn't a great fit. But it's not too bad since it's far closer to 1 than it is to zero.

Let's try another example that is even more problematic.

EXAMPLE 10-2

The following table lists homes by square feet and the corresponding average selling price in Happy Valley, Maine. Find a linear function that fits this data.

SQFT	1200	1500	1750	2000	2250	2500	2750	3000	3500	4000
AveragePrice (thousands)	$135	$142	$156	$165	$170	$220	$225	$275	$300	$450

SOLUTION 10-2

First we enter the data in two arrays:

```
>> sqft = [1200,1500,1750,2000,2250,2500,2750,3000,3500,4000];
>> price = [135,142,156,165,170,220,225,275,300,450];
```

Now let's generate a plot of the data:

```
>> plot(sqft,price,'o'),xlabel('Home SQFT'),
ylabel('Average Selling Price'),...
Title('Average Selling Price by Home SQFT in Happy Valley'),
axis([1200 4000 135 450])
```

The plot is shown in Figure 10-2.

While the price of a 4000 square foot home looks a bit out of the norm, most of the data appears to be roughly on a straight line. Let's try to find out what line best fits this data. The SQFT of the house plays the role of x while average selling price plays the role of y in our attempt to find $y = mx + b$. To use polyfit to find the coefficients we need, we simply pass the data and tell it we are looking for a polynomial of degree one. The call is:

```
>> p = polyfit(sqft,price,1);
```

The function polyfit returns two elements in this case. They are $p(1) = m$ and $p(2) = b$. Let's retrieve those values:

```
>> m = p(1)

m =

    0.1032
```

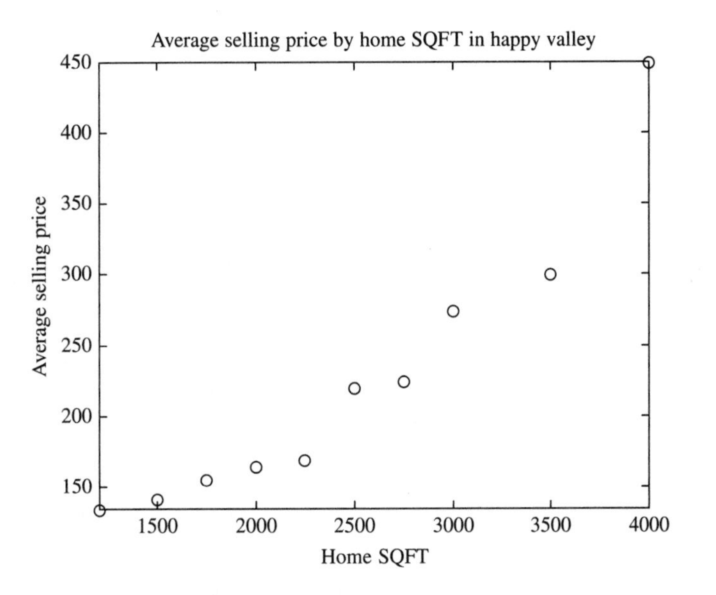

Figure 10-2 The data set used in Example 10-1

```
>> b = p(2)

b =

  -28.4909
```

Next we can plot the fitted line that MATLAB has come up with and compare it with the actual data points. First we generate data for the *x* axis:

```
>> x = [1200:10:4000];
```

Now we use the coefficients we found above to generate *y*:

```
>> y = m*x + b;
```

Finally, we call subplot and then show both curves together on the same graph:

```
>> subplot(2,1,2);

EDU>> plot(x,y,sqft,price,'o'),xlabel('Home SQFT'),
ylabel('Average Selling Price'),...

Title('Average Selling Price by Home SQFT in Happy Valley'),
axis([1200 4000 135 450])
```

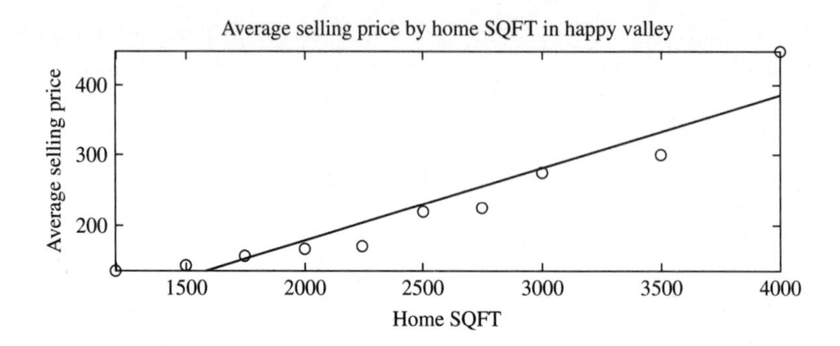

Figure 10-3 Plot showing line MATLAB generated to fit data used in Example 10-2

The data together with the line MATLAB has come up with to fit it is shown in Figure 10-3. While most of the home prices are somewhat close to the fitted line, the homes with the lowest and highest square footage do not. In fact this isn't that great a fit. The polyfit routine has determined that the function that best describes the data is:

$$y = (0.1032)x - 28.4909$$

where x is the square footage of the house and y is the price of the house in thousands. The model would predict that a 1600 sqft house would sell for:

$$y = (0.1032)(1600) - 28.4909 = 136.63$$

or about $136,000. Looking at the data table we can see this is a bit off. Another way we can characterize the goodness of the fit is by calculating a quantity called the *root mean square (RMS) error*. To do this we evaluate the fitted polynomial at the actual data points. This can be done with the following command:

```
>> w = m*sqft + b
```

The data generated are shown here with the original data:

SQFT	1200	1500	1750	2000	2250	2500	2750	3000	3500	4000
Average Price (thousands)	$135	$142	$156	$165	$170	$220	$225	$275	$300	$450
Predicted	$95	$126	$152	$178	$204	$230	$255	$281	$333	$384

You can see that in some cases, the error is not that significant. For instance, with a 2500 SQFT house the difference between the predicted and actual sale price as a percentage is:

$$\frac{|\$230 - \$220|}{\$230} = 0.0455 \approx 4\%$$

While for a 1500 SQFT house:

$$\frac{|\$142 - \$126|}{\$142} = 0.1127 \approx 11\%$$

A realtor probably doesn't demand scientific precision, so a 4% error in an estimate of a sale price might be acceptable with the model, but the 11% error is probably unacceptable. Now let's look at the RMS error to get an overall characterization of the model.

To calculate RMS error, we apply the following formula:

$$\text{RMS error } = \left[\sum_{i=0}^{N} \frac{(T_i - A_i)^2}{N}\right]^{1/2}$$

where T_i are the exact values, A_i are the predictions or approximations of the model, and N is the total number of data points. First let's generate the differences:

```
>> d = price - w;
```

The model has a total of $N = 10$ data points:

```
>> N = 10;
```

To find the RMS error, we need the squares of the differences. We can square each element of the array of differences by writing:

```
>> d2 = d.^2;
```

Now we find the RMS error:

```
>> RMS = sqrt((1/N)*sum(d2))

RMS =

    30.9322
```

A very large error indeed! The RMS error should be less than one. How can we improve the situation? We can try by fitting to a higher order polynomial. Let's try a polynomial of degree two. We can do this with the following steps:

```
>> p = polyfit(sqft,price,2);
```

This time there are three coefficients generated. The function polyfit with the third argument set to two gives us the coefficients for the polynomial:

$$y = p_1x^2 + p_2x + p_3$$

Let's extract these into variables and plot:

```
>> a = p(1);
>> b = p(2);
>> c = p(3);
>> x = [1200:10:4000];
>> y = a*x.^2+ b*x + c;
>> plot(x,y,sqft,price,'o'),xlabel('Home SQFT'),
ylabel('Average Selling Price'),...plot(x,y,sqft,price,'o'),
xlabel('Home SQFT'),ylabel('Average Selling Price'),...
title('Average Selling Price by Home SQFT in Happy Valley'),
axis([1200 4000 135 450])
```

The plot is shown in Figure 10-4. As can be seen, this model is a better fit of the data, although as we'll see it's still far from ideal.

Now let's find the RMS:

```
>> d1 = (w-price).^2;
EDU>> w

w =

  139.5689  142.8071  150.6661  163.2164  180.4581  202.3912
229.0156  260.3314  337.0371  432.5081

>> RMS2 = sqrt((1/N)*sum(d1))

RMS2 =

   15.4324
```

The RMS has been cut in half, a big improvement. Notice by checking the elements in the array *w* that the predicted prices are much closer to the real values. Let's check the *r*-squared value.

```
> M1 = mean(price)

M1 =

  223.8000

>> S = sum((price - M1).^2)

S =

  8.5136e+004

>> w = a*sqft.^2 + b*sqft + c;
>> A = sum((w-price).^2)

A =

  2.3816e+003

>> r2=1-A/S

r2 =

    0.9720
```

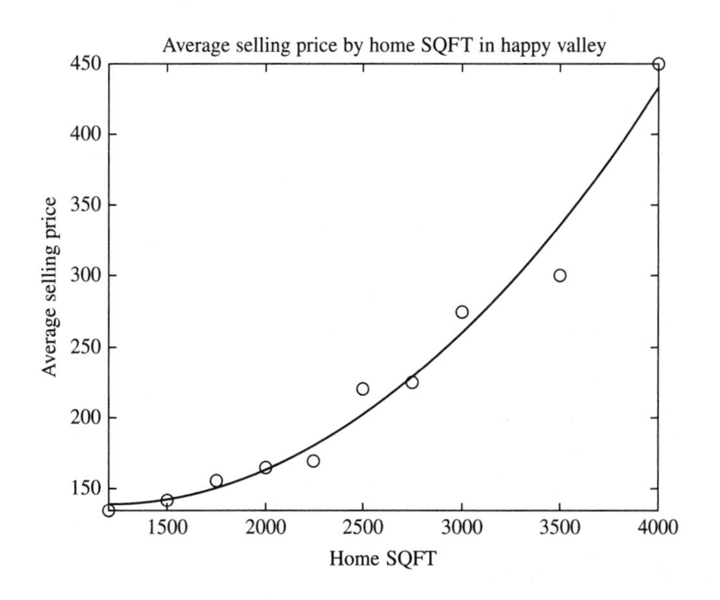

Figure 10-4 Improving our fit by trying a second order polynomial

The *r*-squared value in this case, with the second order polynomial fit, is pretty close to one. So we can take this as a decent model to use.

EXAMPLE 10-3

A block of metal is heated to a temperature of 300 degrees F, and allowed to cool over a period of 7 hours as shown here:

Time (hours)	Temperature (F)
0	300
0.5	281
1.0	261
1.5	244
2.0	228
2.5	214
3.0	202
3.5	191
4.0	181
5.0	164
6.0	151
6.1	149
7.0	141

Try a third order polynomial fit to the data and estimate how good the fit is.

SOLUTION 10-3

We enter the data in MATLAB:

```
>> time = [0,0.5,1.0,1.5,2,2.5,3,3.5,4,5,6,6.1,7];
>>temp = [300,281,261,244,228,214,202,191,181,164,151,149,141];
```

Let's plot the data, which is shown in Figure 10-5.

Now we call polyfit to generate the coefficients for a third order polynomial:

```
>> p = polyfit(time,temp,3);
```

Since the polynomial is third order, our fit function will be of the form:

$$y = p_1 x^3 + p_2 x^2 + p_3 x + p_4$$

Now we extract the coefficients:

```
>> a = p(1); b = p(2); c = p(3); d = p(4);
```

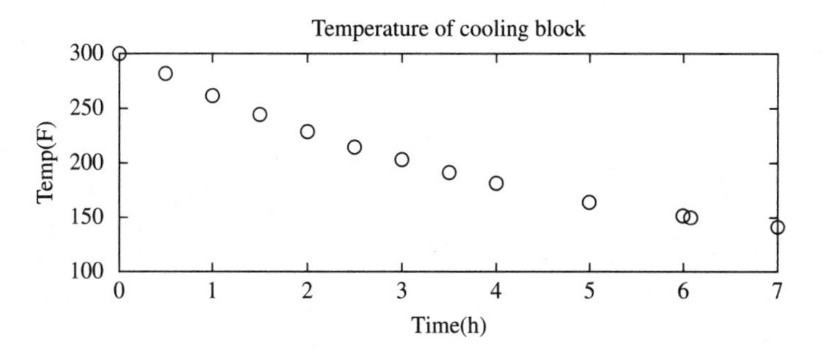

Figure 10-5 A plot of temperatures for a cooling metal block

Let's create data for the time axis:

```
>> t = linspace(0,7);
```

Next define the fit function:

```
>> y = a*t.^3 + b*t.^2 + c*t + d;
```

We will also need an array of data evaluated at the exact time points where the temperature data was collected:

```
>> w = a*time.^3 + b*time.^2 + c*time + d;
```

Let's plot the collected data and the fit curve on the same graph:

```
>> plot(time,temp,'o',t,y),xlabel('time(h)'),ylabel
('temp(F)'),title('Third Order fit for cooling metal block')
```

The result is shown in Figure 10-6. It looks like the third order polynomial has generated a good fit.

Let's compute r^2. First we find the mean of the collected data and calculate S:

```
>> M1 = mean(temp)

M1 =

  208.2308

>> S = sum((temp - M1).^2)

S =

  3.2542e+004
```

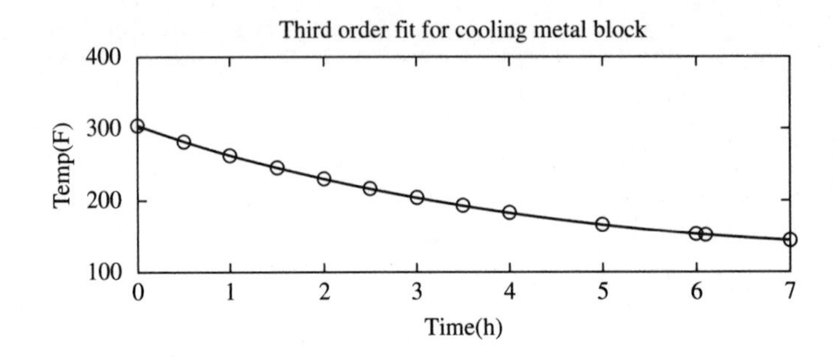

Figure 10-6 Plot of a third order polynomial fit

Now let's find *A*:

```
>> A = sum((w-temp).^2)

A =

   2.79327677455582
```

We find r^2 to be:

```
>> r2 = 1 - A/S

r2 =

    1.0000
```

It looks like we have a perfect fit to the data. However this is misleading and in fact the real function describing this data is:

$$T = 100 + 202.59e^{-0.23t} - 2.59e^{-18.0t}$$

Let's have MATLAB print the *r*-squared value in long format:

```
>> format long
>> r2

r2 =

0.99991416476050
```

Well, the fit is extremely accurate as you can see from the plot. Let's compute the RMS.

```
>> RMS = sqrt(sum((1/N)*(w-temp).^2))

RMS =

   0.46353796413735
```

The RMS is less than one, so we can take it to be a good approximation. Now that we have a function $y(t)$, we can estimate the temperature at various times for which we have no data. For instance, the data is collected out to a time of 7 hours. Let's build the function out to longer times. First we extend the time line:

```
>> t = [0:0.1:15];
```

Regenerate the fit polynomial:

```
>> y = a*t.^3 + b*t.^2 + c*t + d;
```

We can use the find command to ask questions about the data. For instance, when is the temperature less than 80 degrees, so maybe we can risk picking up the metal block?

```
>> find(y < 80)

ans =

   148   149   150   151
```

The command find has returned the array indices for temperatures that satisfy this requirement. We can reference these locations in the array t that contains the times and the array y that contains the temperatures. First we extract the times, adding a quote mark at the end to transpose the data into a column vector:

```
>> A = t(148:151)'

A =

   14.7000
   14.8000
   14.9000
   15.0000
```

Now let's get the corresponding temperatures:

```
>> B = y(148:151)'

B =

   78.5228
   77.0074
   75.4535
   73.8604
```

We can arrange the data into a two column table, with the left column containing the times and the right column the temperatures:

```
>> Table = [A B]

Table =

   14.7000   78.5228
   14.8000   77.0074
   14.9000   75.4535
   15.0000   73.8604
```

So for instance, we see that at 14.9 hours the block is estimated to be at about 75 degrees. Now let's generate a plot of the data from 10 hours to 15 hours. First we find the array index for the time at 10 hours. We had generated the time in increments of 0.1, so we can find it by searching for $t > 9.9$:

```
>> find(t >9.9)

ans =

  Columns 1 through 18

   101   102   103   104   105   106   107   108   109   110
111   112   113   114   115   116   117   118

  Columns 19 through 36

   119   120   121   122   123   124   125   126   127   128
129   130   131   132   133   134   135   136

  Columns 37 through 51

   137   138   139   140   141   142   143   144   145   146
147   148   149   150   151
```

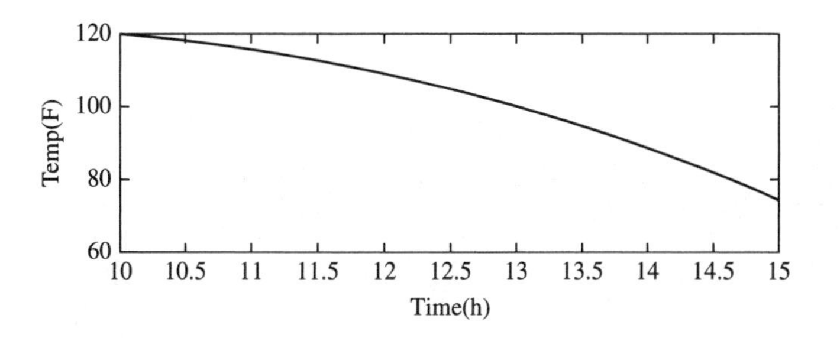

Figure 10-7 A plot of estimated temperatures in a time range well beyond that where the data was collected

Next we create a new array containing these time values:

```
>> T1 = t(101:151);
```

Now let's grab the temperatures over that range and store them:

```
>> TEMP2 = y(101:151);
```

Then we can plot them:

```
>> plot(T1,TEMP2),xlabel('time(h)'),ylabel('Temp(F)'),
axis([10 15 60 120])
```

The result is shown in Figure 10-7. The model predicts that over this 5 hour range the temperature of the metal block will drop more than 40 degrees.

Fitting to an Exponential Function

Other types of fits besides fitting to a polynomial are possible, we briefly mention one. In some cases it may be necessary to fit the data to an exponential function. The fit in this case is:

$$y = b(10)^{mx}$$

where x is the independent variable and y is the dependent variable. Now we define the relations:

$$w = \log_{10} y$$
$$z = x$$

and then fit the data to a line of the form:

$$w = p_1 z + p_2$$

where the coefficients p_1 and p_2 are generated by a call to polyfit. The fit can be generated in MATLAB using the command:

```
p = polyfit(x, log10(y),1)
```

We can find m and b using:

$$m = p_1, \qquad b = 10^{p_2}$$

Quiz

Consider the following set of data. An Olympic weightlifting coach has collected data for maximum number of pounds lifted by the age of the lifter. He believes there is a functional relationship between the two.

Age	Max Weight
15	330
17	370
18	405
19	420
24	550
30	580
35	600
37	580

1. Use MATLAB to find the coefficients m and b for a first order fit to the data.
2. Create an array of ages so that max weight could be estimated by age in 1 year increments in the range of the current team members.
3. Create a function y to implement the first order fit.
4. What does the model predict as the maximum weight lifted by a 17 year old?
5. What is the mean Weight actually lifted?
6. What is the mean weight predicted by the model?
7. To calculate r squared, find S and A.
8. What is r squared for this model?
9. Try a second order fit. What is the function?
10. What is the r-squared value for the second order fit?

CHAPTER 11

Working with Special Functions

In many applications of mathematical physics and engineering the so-called *special functions* rear their ugly head. These are beasts like Bessel functions and the spherical harmonics. In this chapter we will discuss how to work with these types of functions using MATLAB.

Gamma Functions

The *gamma function* that is denoted by $\Gamma(z)$ is defined by the following integral:

$$\Gamma(n) = \int_0^\infty e^{-t} t^{n-1} dt$$

We can see that $\Gamma(1) = 1$ by direct evaluation of the integral:

$$\Gamma(1) = \int_0^\infty e^{-t} t^{1-1} dt = \int_0^\infty e^{-t} dt = -e^{-t}\Big|_0^\infty = 1$$

Using integration by parts, you can show that the gamma function obeys a recursion relation:

$$\Gamma(n+1) = n\Gamma(n)$$

When n is a positive integer, the gamma function turns out to be the factorial function:

$$\Gamma(n+1) = n! \qquad n = 1,2,3,\ldots$$

It can also be shown using the definition in terms of the integral that:

$$\Gamma\left(\frac{1}{2}\right) = \sqrt{\pi}$$

When $0 < x < 1$, the gamma function satisfies:

$$\Gamma(x)\Gamma(1-x) = \frac{\pi}{\sin \pi x}$$

When n is large, the factorial is approximately given by Stirling's formula:

$$n! \sim \sqrt{2\pi n}\, n^n e^{-n}$$

Finally, the gamma function can be used to define Euler's constant, which is:

$$\Gamma'(1) = \int_0^\infty e^{-t} \ln t \, dt = -\gamma = 0.577215\ldots$$

THE GAMMA FUNCTION IN MATLAB

The gamma function of n can be accessed in MATLAB by writing:

```
x = gamma(n)
```

For example, $\Gamma(6) = 5! = 120$ Checking this in MATLAB:

```
>> gamma(6)

ans =

   120
```

So it's easy to use the gamma function in MATLAB to determine factorial values. Let's plot the gamma function. First we define an interval over which to calculate it.

We are going to plot the gamma function for all real values over the interval. The interval is defined as:

```
>> n = linspace(0,5);
```

We'll use plot(x, y) to generate the plot:

```
>> plot(n,gamma(n)),xlabel('n'),ylabel('Gamma(n)'),grid on
```

The result is shown in Figure 11-1. It looks like $\Gamma(n)$ blows up as $n \to 0$, and that it grows very quickly when $n > 5$, as you know from calculating values of the factorial function.

Now let's see how the gamma function behaves for negative values of n. First let's try some integers:

```
>> y = gamma(-1)

y =

   Inf

>> y = gamma(-2)

y =

   Inf
```

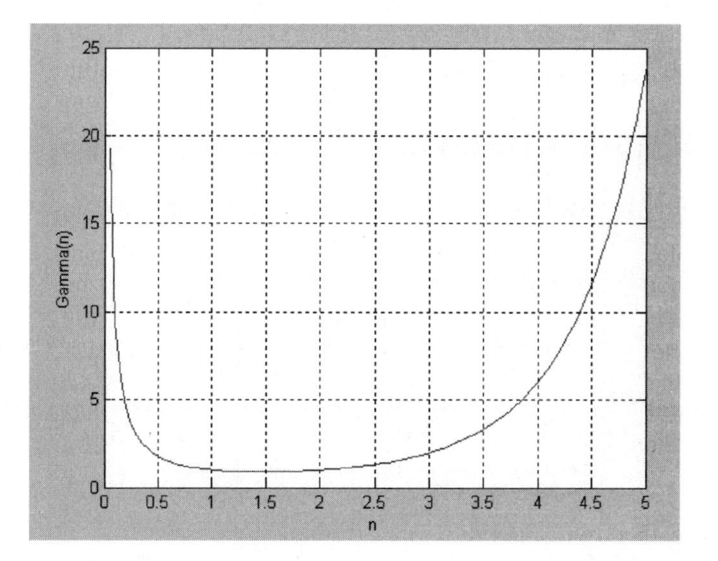

Figure 11-1 A plot of the gamma function for positive values of n

Not looking so good, it appears that the gamma function just blows up for negative argument. But if we try some other values, it looks like this is not always the case. In fact it appears to oscillate between positive and negative values:

```
>> y = gamma(-0.5)

y =

   -3.5449

EDU>> y = gamma(-1.2)

y =

    4.8510

>> y = gamma(-2.3)

y =

   -1.4471
```

Let's try plotting it for negative values. What we find is that the negative integers define asymptotes where $\Gamma(x)$ blows up. This is illustrated in Figure 11-2.

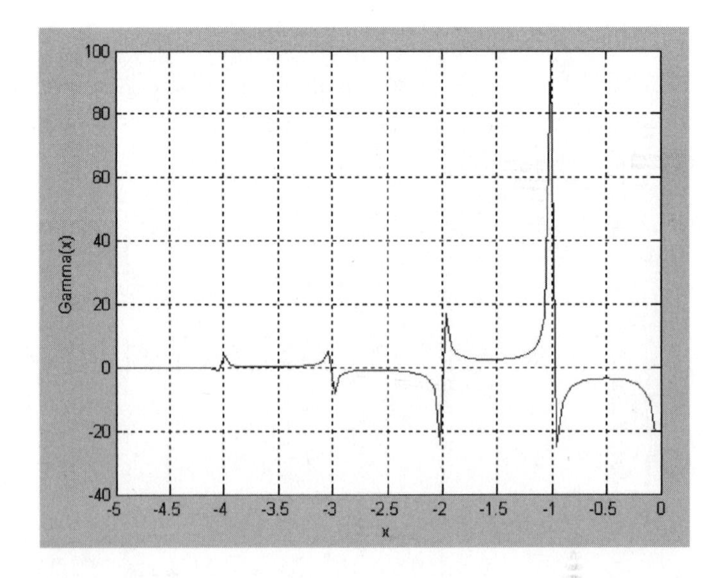

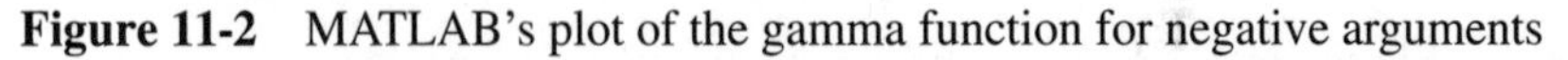
Figure 11-2 MATLAB's plot of the gamma function for negative arguments

Let's tabulate a list of values for the gamma function. We will list values of $\Gamma(x)$ for $1.00 \leq x \leq 2.00$ in increments of 0.1. First we define our list of x values. To display the data in a table format, be sure to include the single quote as shown here (try this example without the tick mark to see how MATLAB displays the data):

```
>> x = (1:0.1:2)';
```

Now we define an array to hold values of the gamma function:

```
>> y = gamma(x);
```

We can generate a tabulated list of values by creating a matrix from these quantities:

```
>> A = [x y]

A =

    1.0000    1.0000
    1.1000    0.9514
    1.2000    0.9182
    1.3000    0.8975
    1.4000    0.8873
    1.5000    0.8862
    1.6000    0.8935
    1.7000    0.9086
    1.8000    0.9314
    1.9000    0.9618
    2.0000    1.0000
```

EXAMPLE 11-1

Use the gamma function to calculate the surface area of a sphere in an arbitrary number of dimensions. Let the radius of the sphere be $r = 1$ and consider the cases of $n = 2, 3, 11$ dimensions.

SOLUTION 11-1

Let the distance measured from the origin or radius of an arbitrary sphere be r. The surface area of a sphere in n dimensions is given by:

$$S_n = r^{n-1} \frac{2\pi^{n/2}}{\Gamma\left(\frac{n}{2}\right)}$$

We can create an inline function in MATLAB to calculate the surface of a sphere:

```
>> surface = @(n,r) r^(n-1)*2*(pi^(n/2))/gamma(n/2)
```

Let's do some quick hand calculations so that we can assure ourselves that the formula works. In two dimensions, a "sphere" is a circle and the surface area is the circumference of the circle:

$$S_2 = r^{2-1}\frac{2\pi^{2/2}}{\Gamma\left(\frac{2}{2}\right)} = r\frac{2\pi}{\Gamma(1)} = 2\pi r$$

since $\Gamma(1) = 1$. If we let $r = 1$, the surface area is:

$$S_2 = 2\pi = 6.2832$$

Calling our function:

```
>> surface(2,1)

ans =

    6.2832
```

Now we move to three dimensions. The surface area of a sphere is:

$$S_3 = r^{3-1}\frac{2\pi^{3/2}}{\Gamma\left(\frac{3}{2}\right)} = r^2\frac{2\pi^{3/2}}{\frac{1}{2}\Gamma\left(\frac{1}{2}\right)} = \frac{4\pi^{3/2}r^2}{\sqrt{\pi}} = 4\pi r^2$$

If $r = 1$, then the surface area is $4\pi = 12.5664$. Checking our function:

```
>> surface(3,1)

ans =

   12.5664
```

Finally, so that string theorists can satisfy their fantasies, we check $n = 11$:

$$S_{11} = r^{11-1}\frac{2\pi^{11/2}}{\Gamma\left(\frac{11}{2}\right)} = \frac{2\pi^{11/2}r^{10}}{\left(\frac{945}{32}\right)\Gamma\left(\frac{1}{2}\right)} = \frac{2\pi^{11/2}r^{10}}{\left(\frac{945}{32}\right)\sqrt{\pi}} = \frac{64}{945}\pi^{11}r^{10}$$

Our function says this is:

```
>> surface(11,1)

ans =

   20.7251
```

QUANTITIES RELATED TO THE GAMMA FUNCTION

MATLAB allows you to calculate the *incomplete gamma function* that is defined by:

$$p(x,n) = \frac{1}{\Gamma(n)} \int_0^x e^{-t} t^{n-1} dt$$

The MATLAB command to evaluate this function is:

```
y = gammainc(x,n)
```

When $x \ll 1$, and $n \ll 1$, the incomplete gamma function satisfies $P(x, n) \approx x^n$. We can check this with a few calculations:

```
>> x = 0.2;
>> n = 0.3;
>> y = x^n

y =

    0.6170

>> z = gammainc(x,n)

z =

    0.6575
```

The approximation is closer for smaller values:

```
>> x = 0.002; n = 0.003; y = x^n

y =

    0.9815
```

```
>> z = gammainc(x,n)

z =

    0.9832
```

Bessel Functions

Bessel's differential equation arises in many scientific and engineering applications. The equation has the form:

$$x^2y'' + xy' + (x^2 - n^2)y = 0$$

Solutions of this equation are:

$$y(x) = A_1J_n(x) + A_2Y_n(x)$$

where A_1 and A_2 are constants determined by the boundary conditions. The functions in the solution are $J_n(x)$ which is a *Bessel function of the first kind* and $Y_n(x)$ which is a *Bessel function of the second kind* or a *Neumann function.*

The Bessel functions of the first kind are defined by the following infinite series, which includes the gamma function as part of the definition:

$$J_n(x) = \sum_{k=0}^{\infty} \frac{(-1)^k (x/2)^{n+2k}}{k!\Gamma(n+k+1)}$$

In MATLAB, the Bessel function of the first kind is implemented by besselj. The call is of the form:

```
y = besselj(n,x)
```

To get a feel for the behavior of the Bessel function of the first kind, let's generate a few plots. We can generate a plot of $J_1(x)$ with the following commands:

```
>> x = [0:0.1:50]; y = besselj(1,x);
>> plot(x,y),xlabel('x'),ylabel('BesselJ(1,x)')
```

The result is shown in Figure 11-3. Notice that as $x \to 0$, $J_1(x)$ is finite, going to 0. We also see that the function has decaying oscillatory behavior.

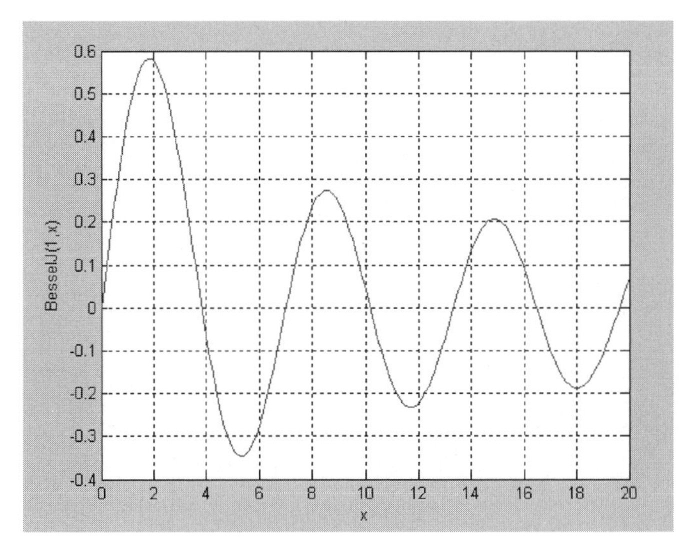

Figure 11-3 A plot of $J_1(x)$

These characteristics apply in general to Bessel functions of the first kind. Let's generate a plot that compares $J_0(x)$, $J_1(x)$, and $J_2(x)$:

```
>> x = [0:0.1:20]; u = besselj(0,x); v = besselj(1,x); w = besselj(2,x);
>> plot(x,u,x,v,'--',x,w,'-.'),xlabel('x'),ylabel('BesselJ(n,x)'),...
grid on, legend('bessel0(x)','bessel1(x)','bessel2(x)')
```

Looking at the plot, shown in Figure 11-4, it appears that the oscillations become smaller as n gets larger.

Bessel functions of the first kind can also be defined for negative integers. They are related to Bessel functions of the first kind for positive integers via:

$$J_{-n}(x) = (-1)^n J_n(x)$$

Let's do some symbolic computation to find the derivative of a Bessel function:

```
>> syms n x y
>> diff(besselj(n,x))

ans =

-besselj(n+1,x)+n/x*besselj(n,x)
```

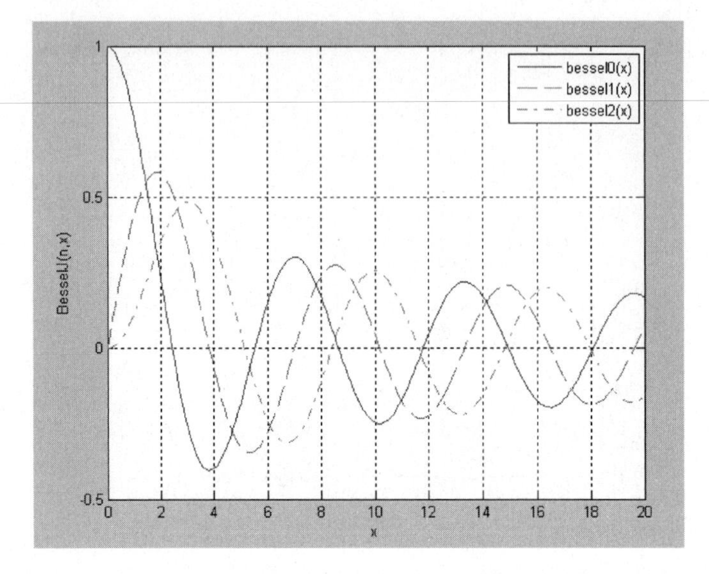

Figure 11-4 A plot of $J_0(x)$, $J_1(x)$, and $J_2(x)$

MATLAB has calculated the relation:

$$J_n'(x) = \frac{n}{x} J_n(x) - J_{n+1}(x)$$

When integrals involving Bessel functions are encountered, MATLAB can also be used. For example:

```
>> int(x^n*besselj(n-1,x))

ans =

x^n*besselj(n,x)
```

That is, modulo a constant of integration:

$$\int x^n J_{n-1}(x) dx = x^n J_n(x)$$

Let's compute the first five terms of the Taylor series expansions of $J_0(x)$ and $J_1(x)$:

```
>> taylor(besselj(0,x),5)

ans =

1-1/4*x^2+1/64*x^4

>> taylor(besselj(1,x),5)

ans =

1/2*x-1/16*x^3
```

MATLAB also has the other Bessel functions built in. Bessel functions of the second kind are implemented with bessely(n, x). In Figure 11-5, we show a plot of the first three Bessel functions of the second kind generated with MATLAB.

We can also implement another type of Bessel function in MATLAB, the *Hankel* function. These can be utilized by calling bessel$h(nu, k, z)$. There are two types of Hankel functions (first kind and second kind), where the kind is denoted by k in MATLAB. If we leave the k out of the argument and call bessel$h(nu, z)$ MATLAB defaults to the Hankel function of the first kind.

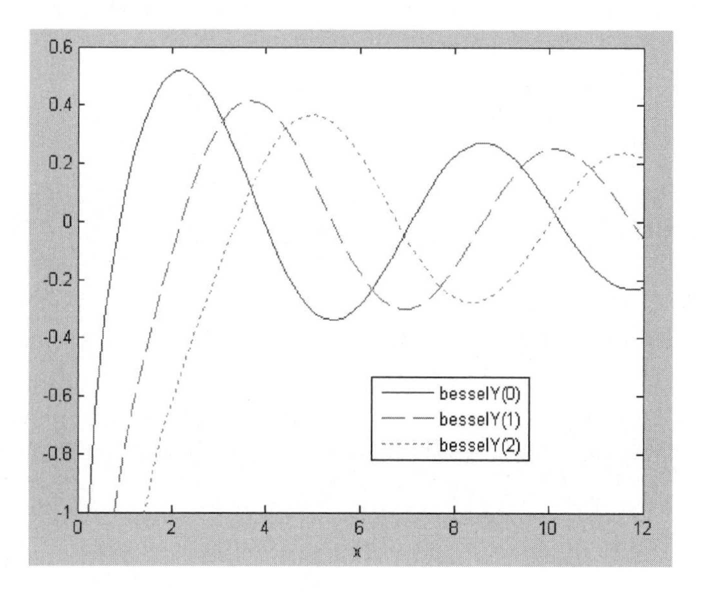

Figure 11-5 Plotting Bessel functions of the second kind

EXAMPLE 11-2

A cylindrical traveling wave can be shown to be described by:

$$\Psi(r,t) = H_0^{(1)}(kr)e^{-i\omega t}$$

where we take the real part. Implement the real part of this function in MATLAB and plot as a function of r (in meters) for $t = 0$, $\pi/2\omega$, $3\pi/4\omega$. Assume that $k = 500\ \text{cm}^{-1}$ over $0 \le r \le 10$ m.

SOLUTION 11-2

First we convert k into the proper units:

$$k = 500\,\text{cm}^{-1}\left(\frac{100\,\text{cm}}{\text{m}}\right) = 50000\,\text{m}^{-1}$$

Let's generate some plots to see how k impacts the behavior of the real part of the Hankel function. First we generate an array for the range of values over which we want to plot:

```
>> r = linspace(0,10);
```

Let's suppose that k were unity. In this case we could define the Hankel function as:

```
>> u = besselh(0,r);
```

You can evaluate the Hankel function at different points to see that it's complex. For example:

```
>> besselh(0,2)

ans =

   0.2239 + 0.5104i
```

This indicates that for the function in question to be physical, we have to take the real part, a common practice when studying traveling electromagnetic waves, for example. So we plot the real part of this function:

```
>> plot(r,real(u)),xlabel('r'),ylabel('Hankel(0,r)')
```

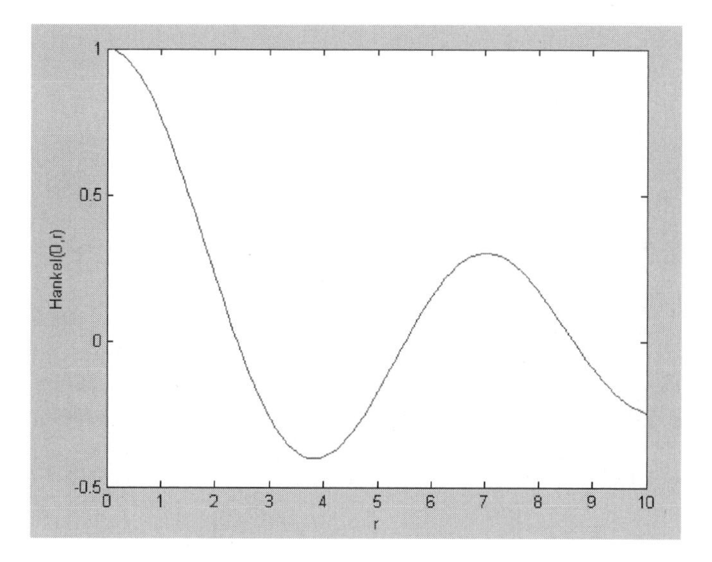

Figure 11-6 A plot of the real part of $H_0^{(1)}(r)$ for $0 \leq r \leq 10$

The result, shown in Figure 11-6, indicates that this function is a decaying oscillatory function. Not surprising since this is a type of Bessel function and we have already seen that Bessel functions are decaying oscillatory functions!

Now let's add in the wave number k. The code above becomes:

```
>> r = linspace(0,10);
>> k = 50000;
>> w = besselh(0,k*r);
>> plot(r,real(w)),xlabel('r'),ylabel('Hankel(0,kr)')
```

Two items are immediately apparent by looking at the graph, shown in Figure 11-7. The first is that the function appears to oscillate quite a bit more rapidly over the distance r. Secondly, we notice that the amplitude is much smaller in this case.

Now let's take a look at the function at different times. First considering $t = 0$, we have:

$$\psi(r,0) = H_0^{(1)}(kr)$$

This is what we have in Figure 11-7. Now let's let $t = \pi/2\omega$.

$$\psi(r,\pi/2\omega) = H_0^{(1)}(kr)e^{-i\omega(\pi/2\omega)} = H_0^{(1)}(kr)e^{-i\pi/2} = -iH_0^{(1)}(kr)$$

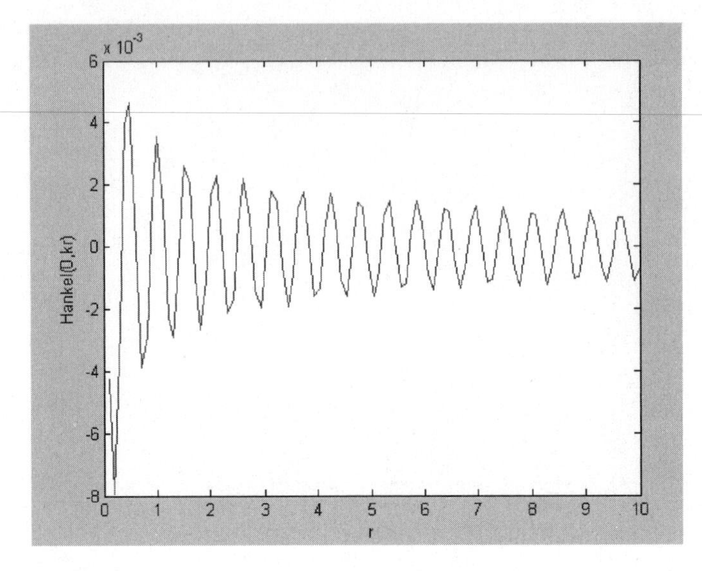

Figure 11-7 Plot of the Hankel function with $k = 50{,}000$

To obtain this result we used Euler's formula:

$$e^{i\theta} = \cos\theta + i\sin\theta$$

In Figure 11-8, we show a plot of the real part of this function (dashed line) on the same graph with the function at $t = 0$.

There doesn't seem to be much difference here. The waveform is a bit displaced. The function then describes a decaying wave as r gets larger.

For the final time, we create the function:

```
>> s = besselh(0,k*r)*(cos(3*pi/4)-i*sin(3*pi/4));
```

The plot is shown in Figure 11-9.

A check of the Hankel function shows it cannot be evaluated at zero:

```
>> besselh(0,0)

ans =

       NaN +     NaNi
```

MATLAB uses NaN to represent "not a number."

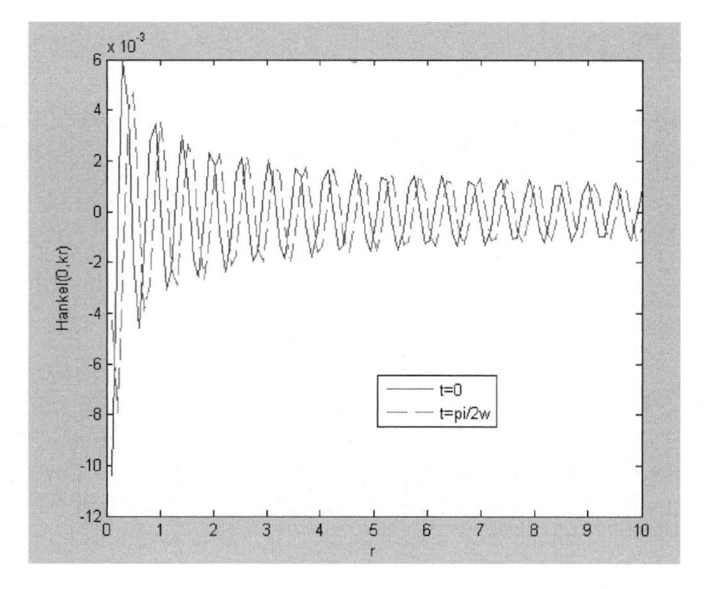

Figure 11-8 A plot of the function at $t = 0$ and $t = \pi/2\omega$

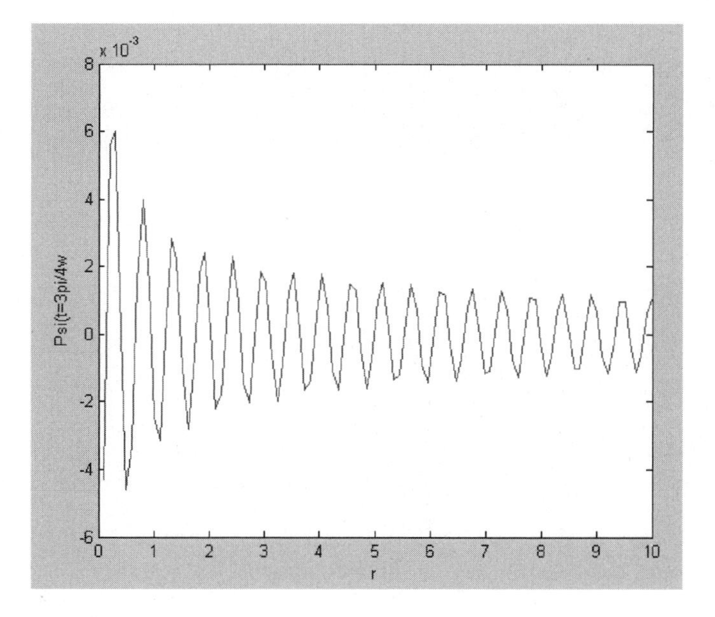

Figure 11-9 A plot showing the real part of $\Psi(r, t) = H_0^{(1)}(kr)e^{-i\omega t}$ at $t = 3\pi/4\omega$

The Beta Function

The *Beta function* takes two arguments and is defined in terms of the following integral:

$$B(m,n) = \int_0^1 x^{m-1}(1-x)^{n-1}\,dx$$

This integral converges if $m, n > 0$. To evaluate the Beta function in MATLAB, we use:

```
x = beta(m,n)
```

We can use MATLAB to generate tables of values for the Beta function. If $m = 1$, we find that the first ten values of beta(1, n) are:

```
>> x = (1:1:10)'; y = beta(1,x); A = [x y]

A =

    1.0000    1.0000
    2.0000    0.5000
    3.0000    0.3333
    4.0000    0.2500
    5.0000    0.2000
    6.0000    0.1667
    7.0000    0.1429
    8.0000    0.1250
    9.0000    0.1111
   10.0000    0.1000
```

What good is the Beta function? It turns out that the Beta function can be used to evaluate integrals involving products of $\sin\theta$ and $\cos\theta$.

EXAMPLE 11-3

Determine the value of

$$\int_0^{\pi/2} \sin^5\theta\cos^3\theta\,d\theta$$

SOLUTION 11-3

To do this integral we could try to dig out some tricks from a calculus book or leave on an endless journey of integration by parts. Instead we can use the relation:

$$\int_0^{\pi/2} \sin^{2m-1}\theta \cos^{2n-1}\theta \, d\theta = \frac{1}{2} B(m,n)$$

Hence:

$$\int_0^{\pi/2} \sin^5\theta \cos^3\theta \, d\theta = \frac{1}{2} B(3,2)$$

Calling on MATLAB for the answer:

```
>> (0.5)*beta(3,2)

ans =

    0.0417
```

$$\Rightarrow \int_0^{\pi/2} \sin^5\theta \cos^3\theta \, d\theta = 0.0417$$

EXAMPLE 11-4

Find

$$\int_0^{\pi/2} \sin^{19}\theta \, d\theta$$

SOLUTION 11-4

This time $m = 10$, $n = 1/2$ and so the integral is given by:

```
>> (1/2)*beta(10,1/2)

ans =

    0.2838
```

EXAMPLE 11-5
Evaluate

$$\int_0^{\pi/2} \sqrt{\tan\theta}\, d\theta$$

SOLUTION 11-5
Some minor manipulation will turn this into the form that can be evaluated using the Beta function:

$$\int_0^{\pi/2} \sqrt{\tan\theta}\, d\theta = \int_0^{\pi/2} \sqrt{\frac{\sin\theta}{\cos\theta}}\, d\theta = \int_0^{\pi/2} \sin^{1/2}\theta \cos^{-1/2}\theta\, d\theta$$

So we have:

$$\begin{aligned} 2m - 1 &= 1/2, \quad &\Rightarrow m = 3/4 \\ 2n - 1 &= -1/2, \quad &\Rightarrow n = 1/4 \end{aligned}$$

So the integral evaluates to $1/2B(3/4,\ 1/4)$ which is:

```
>> 0.5*beta(3/4,1/4)

ans =

    2.2214
```

This happens to be $\pi/\sqrt{2}$.

Special Integrals

There are many "special integrals" in mathematical physics. In this section, we will mention a few of them and see how to calculate with them in MATLAB. The first we consider is the *exponential integral*:

$$E_i(x) = \int_x^{\infty} \frac{e^{-u}}{u}\, du$$

This function is implemented in MATLAB with the following syntax:

```
y = expint(x)
```

Here we generate a tabulated list of values. Note that expint(0) = inf.

```
>> x = (0.1:0.1:2)'; y = expint(x); A = [x y]

A =

    0.1000    1.8229
    0.2000    1.2227
    0.3000    0.9057
    0.4000    0.7024
    0.5000    0.5598
    0.6000    0.4544
    0.7000    0.3738
    0.8000    0.3106
    0.9000    0.2602
    1.0000    0.2194
    1.1000    0.1860
    1.2000    0.1584
    1.3000    0.1355
    1.4000    0.1162
    1.5000    0.1000
    1.6000    0.0863
    1.7000    0.0747
    1.8000    0.0647
    1.9000    0.0562
    2.0000    0.0489
```

Many other special functions can be numerically evaluated using the mfun command. Calling on MATLAB help, we see a listing of functions that can be numerically evaluated using this command:

```
>> help mfunlist
```

mfunlist special functions for mfun.

The following special functions are listed in alphabetical order according to the third column. n denotes an integer argument, x denotes a real argument, and z denotes a complex argument. For more detailed descriptions of the functions, including any argument restrictions, see the Reference Manual, or use MHELP.

```
bernoulli     n       Bernoulli numbers
bernoulli     n,z     Bernoulli polynomials
BesselI       x1,x    Bessel function of the first kind
BesselJ       x1,x    Bessel function of the first kind
BesselK       x1,x    Bessel function of the second kind
BesselY       x1,x    Bessel function of the second kind
Beta          z1,z2   Beta function
```

```
binomial        x1,x2   Binomial coefficients
EllipticF -     z,k     Incomplete elliptic integral, first kind
EllipticK -     k       Complete elliptic integral, first kind
EllipticCK      k       Complementary complete integral, first kind
EllipticE -     k       Complete elliptic integrals, second kind
EllipticE -     z,k     Incomplete elliptic integrals, second kind
EllipticCE      k       Complementary complete elliptic integral, second kind
EllipticPi -    nu,k    Complete elliptic integrals, third kind
EllipticPi -    z,nu,k  Incomplete elliptic integrals, third kind
EllipticCPi     nu,k    Complementary complete elliptic integral, third kind
erfc            z       Complementary error function
erfc            n,z     Complementary error function's iterated integrals
Ci              z       Cosine integral
dawson          x       Dawson's integral
Psi             z       Digamma function
dilog           x       Dilogarithm integral
erf             z       Error function
euler           n       Euler numbers
euler           n,z     Euler polynomials
Ei              x       Exponential integral
Ei              n,z     Exponential integral
FresnelC        x       Fresnel cosine integral
FresnelS        x       Fresnel sine integral
GAMM            z       Gamma function
harmonic        n       Harmonic function
Chi             z       Hyperbolic cosine integral
Shi             z       Hyperbolic sine integral
GAMMA           z1,z2   Incomplete gamma function
W               z       Lambert's w function
W               n,z     Lambert's w function
lnGAMMA         z       Logarithm of the gamma function
Li              x       Logarithmic integral
Psi             n,z     Polygamma function
Ssi             z       Shifted sine integral
Si              z       Sine integral
Zeta            z       (Riemann) zeta function
Zeta            n,z     (Riemann) zeta function
Zeta            n,z,x   (Riemann) zeta function
  Orthogonal Polynomials (Extended Symbolic Math Toolbox only)
T               n,x     Chebyshev of the first kind
U               n,x     Chebyshev of the second kind
G               n,x1,x  Gegenbauer
H               n,x     Hermite
P               n,x1,   Jacobi
                x2,x
L               n,x     Laguerre
L               n,x1,x  Generalized laguerre
P               n,x     Legendre
```

See also mfun, mhelp.

To see how to use this function, we consider the Riemann zeta function. It is given by the series:

$$\zeta(z) = \sum_{n=1}^{\infty} \frac{1}{n^z}$$

The argument z is a complex number such that Re(z) > 1. The evaluation of the Zeta function at real integers produces some interesting results, as we see from the first few values. For $z = 1$, we obtain the harmonic series, which blows up:

$$\zeta(1) = 1 + \frac{1}{2} + \frac{1}{3} + \frac{1}{4} + \cdots = \propto$$

Moving along, we get some interesting relations:

$$\zeta(2) = 1 + \frac{1}{2^2} + \frac{1}{3^2} + \cdots = \frac{\pi^2}{6} = 1.645$$

$$\zeta(3) = 1 + \frac{1}{2^3} + \frac{1}{3^3} + \cdots = 1.202$$

In fact whenever the argument is even, say m is an even number, then the Zeta function is proportional to:

$$\zeta(m) \cdot \pi^m$$

To evaluate the Riemann Zeta function in MATLAB, we write w = mfun('Zeta', z). For example, we can estimate the value of π by calculating $\zeta(2)$:

```
w = mfun('Zeta',2)

w =

    1.6449

>> my_pi = sqrt(6*w)

my_pi =

    3.1416
```

Let's plot the Zeta function for real positive argument *x*. First we define our interval:

```
>> x = linspace(0,10);
```

Next, we define the Zeta function and plot:

```
>> w = mfun('Zeta',x);

>> plot(x,w),xlabel('x'),title('Zeta(x)'),axis([0 10 -10 10])
```

The plot is shown in Figure 11-10. As we saw earlier, the Zeta function blows up at $x = 1$, and this is indicated in the plot. For $x > 1$ it quickly decays. We can verify that it converges to one by checking a few values:

```
>> w = mfun('Zeta',1000)

w =

     1

>> w = mfun('Zeta',1000000)

w =

     1
```

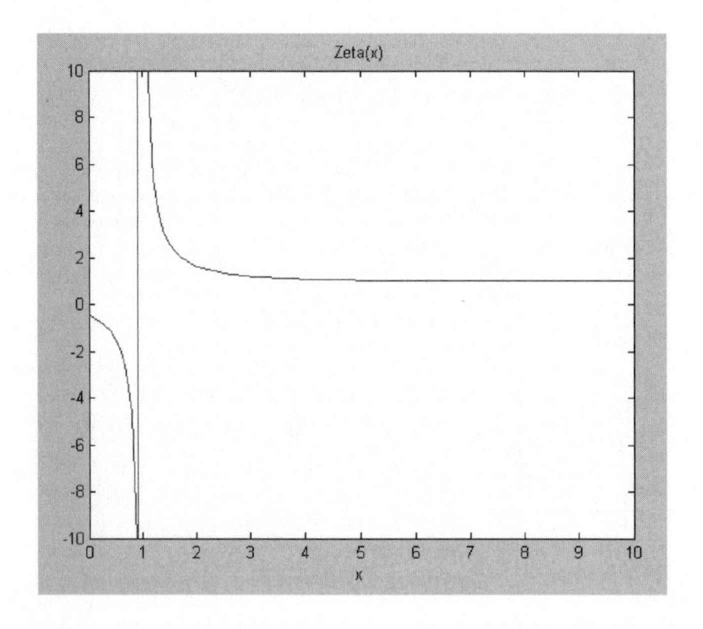

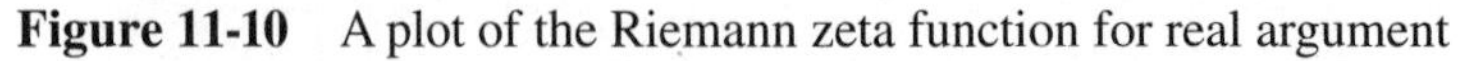
Figure 11-10 A plot of the Riemann zeta function for real argument

Legendre Functions

Legendre's differential equation is given by:

$$(1-x^2)y'' - 2xy' + n(n+1)y = 0$$

where n is a nonnegative integer. The solution of this equation can be written as:

$$y = a_1P_n(x) + a_2Q_n(x)$$

The $P_n(x)$ are polynomials called Legendre polynomials. They can be derived using Rodrigue's formula:

$$P_n(x) = \frac{1}{2^n n!}\frac{d^n}{dx^n}(x^2-1)^n$$

Legendre's associated differential equation is given by:

$$(1-x^2)y'' - 2xy' + \left[n(n+1) - \frac{m^2}{1-x^2}\right]y = 0$$

Here m is also a nonnegative integer. Solutions of this equation have the form:

$$y = a_1P_n^m(x) + a_2Q_n^m(x)$$

Here $P_n^m(x)$ are the associated Legendre functions of the first kind, and $Q_n^m(x)$ are the associated Legendre functions of the second kind. $P_n^m(x) = 0$ if $m > n$. The associated Legendre function $P_n^m(x)$ can be evaluated in MATLAB using the following command:

```
p = legendre(n,x)
```

The associated Legendre function is valid only for $-1 \le x \le 1$. The function legendre(n, x) calculates the associated l Legendre functions $P_n^0(x)$, $P_n^1(x)$,..., $P_n^n(x)$. This can be used for numerical evaluation of the associated Legendre functions.

For example, suppose that we want to evaluate the case $n = 1$. This tells us that $m = 0$ or $m = 1$. First let's pick some regularly spaced points over $-1 \le x \le 1$:

```
>> x = (-1:0.5:1)

x =

   -1.0000   -0.5000         0    0.5000    1.0000
```

Now we evaluate the function for $n = 1$

```
>> p = legendre(1,x)
```

The data is presented in the following way:

```
p =

   -1.0000   -0.5000         0    0.5000    1.0000

         0   -0.8660   -1.0000   -0.8660         0
```

The value of m labels the rows, while the value of x labels the columns. We will put this in a formal table so that you can understand how this works:

m	x = -1.0000	-0.5000	0	0.5000	1.0000
0	-1.0000	-0.5000	0	0.5000	1.000
1	0	-0.8660	-1.0000	-0.8660	0

Hence $P_1^0(-1) = -1$, $P_1^1(-0.5) = -0.866$, for example. Plotting these functions gives some interesting results. In Figure 11-11, we show a plot of legendre(1, x) . This includes curves for $m = 0$ and $m = 1$. The straight line is $P_1^0(x)$ and the curved line is $P_1^1(x)$. The code used to generate the plot is as follows. First we define the interval:

```
>> x = linspace(-1,1);
```

Next we build some associated Legendre functions:

```
>> p1 = legendre(1,x); p2 = legendre(2,x); p3 = legendre(3,x);
p4 = legendre(4,x);
```

To plot the first one, we type:

```
>> plot(x,p1),xlabel('x'),ylabel('p1')
```

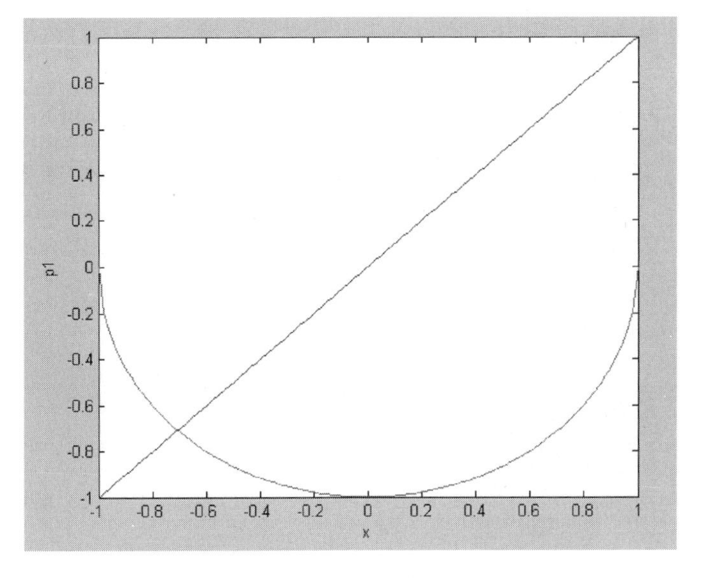

Figure 11-11 A plot of $P_1^0(x)$ and $P_1^1(x)$

Now let's generate plots for legendre(2, *x*). In this case $n = 2$ which means that $m = 0, 1, 2$. The result is shown in Figure 11-12. First, let's see how to extract each of the functions. This is done using standard MATLAB array addressing. As we saw earlier, each row of legendre(*n*, *x*) represents the function for a value of *m*. In the case of legendre(2, *x*), row 1 is $P_2^0(x)$, row 2 is $P_2^1(x)$, and row 3 is $P_2^2(x)$. Now let's extract the data into separate arrays. By typing $b = A(i, :)$ we tell MATLAB to take the ith row with all columns (indicated by the :) and place the data into the array *b*. In this case we write:

```
>> f = p2(1,:);
>> g = p2(2,:);
>> h = p2(3,:);
```

Now we can plot all three functions on the same graph, identifying each individual curve:

```
>> plot(x,f,x,g,'--',x,h,':'),xlabel('x'),ylabel('p2'),
legend('p(2,0)','p(2,1)','p(2,2)')
```

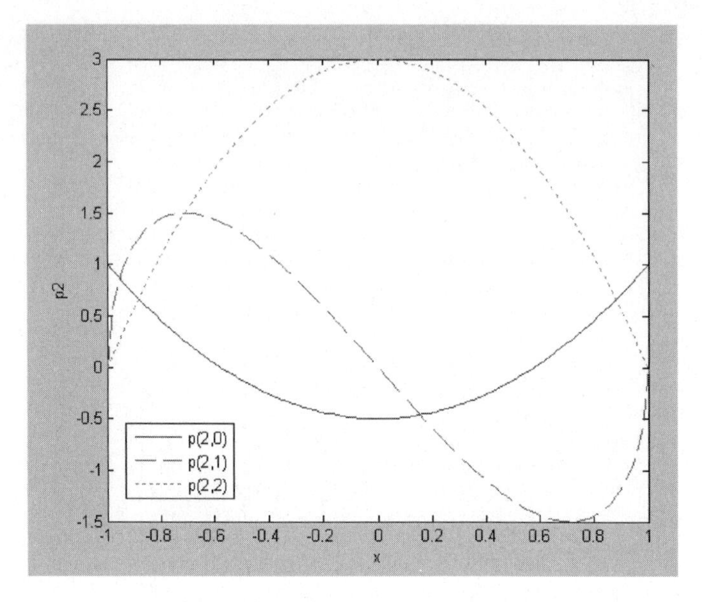

Figure 11-12 Plot of the three functions associated with Legendre(2, x)

Airy Functions

Airy functions are solutions to a differential equation of the form:

$$\frac{d^2\Psi}{dx^2} - x\Psi = 0$$

In physics, Airy functions arise in the study of quantum mechanics. We denote Airy functions by $Ai(z)$. When the argument of the Airy function is real, it can be written as:

$$Ai(x) = \frac{1}{\pi}\int_0^\infty \cos\left(\frac{t^3}{3} + xt\right)dt$$

In MATLAB, w = airy(z) can be used to evaluate $Ai(z)$. For real argument, at $x = 0$ the Airy function can be written in terms of the gamma function in the following way:

$$Ai(0) = \frac{1}{3^{2/3}\Gamma(2/3)}$$

As $x \to \infty$, the Airy function assumes the asymptotic form:

$$Ai(x) \sim \frac{1}{2\sqrt{\pi}} \frac{e^{-2/3x^{3/2}}}{x^{1/4}}$$

As $x \to -\infty$, we have:

$$Ai(x) \sim \frac{1}{\sqrt{\pi}} \frac{\sin\left(\frac{2}{3}x^{3/2} - \frac{\pi}{4}\right)}{x^{1/4}}$$

Let's use MATLAB to plot the Airy function. First we choose an interval that will bring out the main features of the function:

```
>> x = linspace(-10,5);
```

Now we construct an array of data evaluated for the Airy function:

```
>> y = airy(x);
```

and plot it:

```
>> plot(x,y),xlabel('x'),ylabel('Ai(x)'),grid on
```

CAUTION *Imaginary parts of complex X and/or Y arguments ignored.*

The result is shown in Figure 11-13. You can see the oscillatory behavior for negative x, and the exponential decay for positive x, which is consistent with the asymptotic formulas listed earlier.

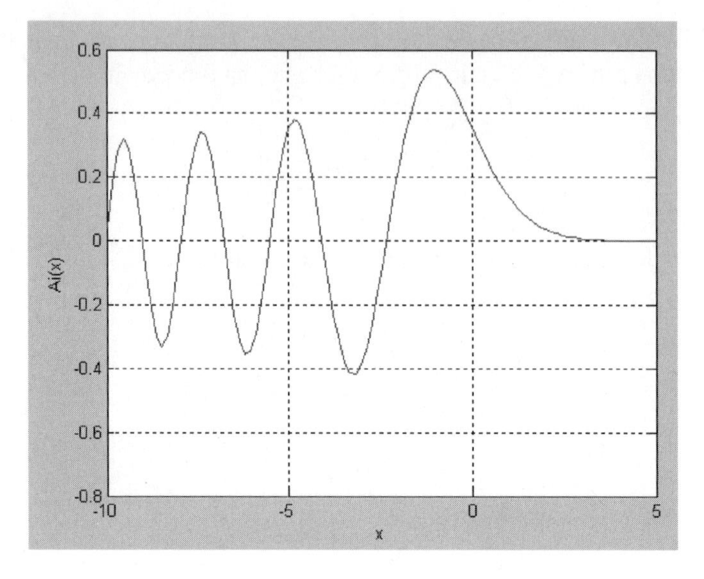

Figure 11-13 A plot of the Airy function

Quiz

1. Use the gamma function to calculate 17.
2. Use the gamma function in MATLAB to evaluate $\int_0^\infty 5^{-2z^2} dz$ (use substitution techniques to rewrite the integral in terms of the gamma function).
3. Create a tabulated list of values of the gamma function for $1 \le x \le 2$ in increments of 0.1.
4. Using MATLAB, find $\int_0^1 x\, J_n(2x)\, J_n(x) dx$.
5. The volume of a sphere in n dimensions is:

$$V_n = r^n \frac{\pi^{n/2}}{\Gamma(1+n/2)}$$

Write a MATLAB function to implement this calculation and determine the volume of a sphere with a radius of 3 m in five dimensions.

6. Calculate the following integral:

$$\int_0^{\pi/2} \sin^{13}\theta \cos^{15}\theta \, d\theta$$

7. Plot the associated Legendre polynomial $P_3^2(x)$.
8. Calculate the integral:

$$\int_0^{\pi/2} \cos^{17}\theta \, d\theta$$

9. Create a table of values of the Airy function for $0 \le x \le 5$ with an increment of 0.5.
10. Check the relation between the Airy function and the gamma function at $x = 0$.

APPENDIX

Bibliography and References

MATLAB online help Version 7.1, The Math Works Inc., Natick, Massachusetts, 2005.

E. W. Nelson, C. L. Best, and W. G. McLean, *"Schaum's Outlines Engineering Mechanics: Statics and Dynamics,"* 5th Ed., McGraw-Hill, New York, 1998.

William J. Palm, "*Introduction to MATLAB 7 For Engineers*," 2d Ed., McGraw-Hill, New York, 2005.

David A. Sanchez, Richard C. Allen, Walter T. Kyner, "*Differential Equations*," 2d Ed., Addison Wesley, New York, 1988.

Francis Scheid, "*Schaum's Outline of Numerical Analysis*," 2d Ed., McGraw-Hill, New York, 1988.

S. K. Stein, "*Calculus and Analytic Geometry*," 4th Ed., McGraw-Hill, New York, 1987.

Final Exam

Use MATLAB to compute the following quantities.

1. $4\frac{7}{8} - 2^3 + 3(4^3)$
2. $5(12^{3/4} - 2)$
3. $36^{0.78}$
4. Using MATLAB, find:

$$\frac{d}{dx}(x^2 \cos(\pi x))$$

5. Find the first derivative of $f(x) = \frac{2x}{x^2 - 9}$.
6. Find the second derivative of $x \sin(x)$.
7. Using MATLAB, find the binomial expansion of $(1 + x)^4$.
8. What are the critical points of $f(x) = 2x^3 - 3x^2$?

9. Find the limit: $\lim_{x \to 1} \frac{x^3 - 1}{x^2 - 1}$.
10. Find the limit: $\lim_{x \to 0} 3^{1/x}$.
11. Find the limit: $\lim_{x \to 0^+} 3^{1/x}$.
12. Find the limit $\lim_{x \to 0^-} 3^{1/x}$.
13. Calculate the derivative of $\sqrt[4]{x}$ and evaluate at $x = 67$.
14. Substitute $x = 1.1$ and $x = 2.7$ into the function $x^2 - 12x + 4$.
15. What is the third derivative of $\left(\sqrt{x} + 2\right)^{99}$.
16. What MATLAB function can be used to find the roots of an equation?
17. We use MATLAB to solve an equation $2x + 3 = 0$. What is the correct function to call and what is the syntax:
    (a) find('2 * x + 3')
    (b) roots('2 * x + 3')
    (c) solve('2 * x + 3')
    (d) solve('2 * x + 3', 0)
18. We use MATLAB to solve an equation $2x + 3 = 1$. What is the correct function to call and what is the syntax:
    (a) roots('2 * x + 3 = 1')
    (b) solve('2 * x + 3 = 1')
    (c) solve('2 * x + 3', 1)
    (d) roots('2 * x + 3', 1)
19. Find the roots of $x^2 - 5x + 9 = 0$.
20. Use MATLAB to factor $x^3 - 64$.
21. To calculate the limit $\lim_{x \to 2^+} \frac{x + 3}{x - 4}$ the correct MATLAB call is:
    (a) limit((x + 3)/(x – 4), 2, +)
    (b) limit((x + 3)/(x – 4)), 2,+
    (c) limit((x + 3)/(x – 4), 2, 'right')
    (d) limit((x + 3)/(x – 4), 2), right

For questions 22–24, let $f(x) = x^2$ over [1, 3].

22. What is the minimum?
23. What is the maximum?

24. Find the average value of the function.
25. If MATLAB prints a function *f as*

    $f =$

    $x \wedge 3 - 3 * x \wedge 2 - 4 * x + 2$

    What function can we use to have it display as $x^3 - 3x^2 - 4x + 2$?
26. To generate a set of uniformly spaced points for $0 \le x \le 10$ you can write:
    (a) *x* = linspace(0:10);
    (b) *x* = linspace(0:0.1:10);
    (c) *x* = linspace(0, 10, '*u*');
    (d) *x* = linspace(0, 10);
27. You want to plot two curves *y*1 and *y*2 against *x*. To plot *y*1 as a red line and *y*2 as a blue line, the correct command is:
    (a) plot(*x*, *y*1, '*r*', *x*, *y*2, '*b*')
    (b) plot(*x*, *y*1, '*r*', *y*2, '*b*')
    (c) plot(*x*, *y*1, "*r*", *y*2, "*b*")
    (d) plot(*x*, *y*1, *x*, *y*2), color('*r*', '*b*')
28. You want to plot a curve as a dashed blue line. The correct command is:
    (a) plot(*x*, *y*, '*b*–')
    (b) plot(*x*, *y*, '*b*–')
    (c) plot(*x*, *y*, '*b*', '–')
29. To add a title to a graph, the correct command is:
    (a) plot(*x*, *y*, 'title–>'Plot of Sin(*x*)')
    (b) plot(*x*, *y*, 'Plot of Sin(*x*)')
    (c) plot(*x*, *y*), title('Plot of Sin(*x*)')
    (d) plot(*x*, *y*), Title('Plot of Sin(*x*)')
30. To plot a curve as a red dotted line, the correct command is:
    (a) plot(*x*, *y*, '*r*', ':')
    (b) plot(*x*, *y*, '*r*:')
    (c) plot(*x*, *y*, '*r*.')

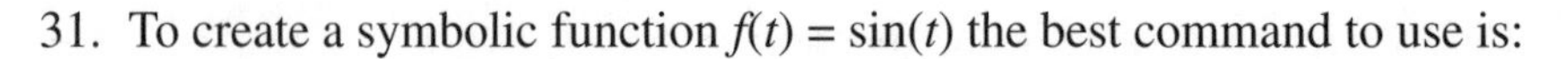

31. To create a symbolic function $f(t) = \sin(t)$ the best command to use is:
    (a) syms t; $f = \sin(t)$;
    (b) syms t; $f = \sin(\text{'}t\text{'})$;
    (c) syms t; $f =$ 'sin(t)';
    (d) *a* or *c*
32. To plot a symbolic function f you call:
    (a) quickplot(f)
    (b) symplot(f)
    (c) ezplot(f)
    (d) plot(f, 'symbolic')
33. You want to use *ez*plot to generate a graph of a function f and its second derivative on the *same graph*. The correct command is:
    (a) ezplot(f, diff(f, 2))
    (b) subplot(1, 2, 1); ezplot(f) subplot(1, 2, 2); ezplot(diff(f, 2))
    (c) ezplot(f); hold on ezplot(diff(f, 2))
    (d) ezplot(f); hold; ezplot(diff(f, 2));
34. You want to use *ez*plot to generate a graph of a function f and its second derivative in side-by-side plots. The correct command is:
    (a) subplot(1, 2, 1); ezplot(f) subplot(1, 2, 2); ezplot(diff(f, 2))
    (b) ezplot(subplot(1, f), subplot(2, diff(f, 2))
    (c) ezplot(f); hold on ezplot(diff(f, 2))
35. You have created a symbolic function f. To display its third derivative, you can write:
    (a) f'''
    (b) f''' or diff(f, 3)
    (c) derivative(f, 3)
    (d) diff(f, 3)
36. The polyfit(x, y, n) function returns
    (a) A symbolic polynomial function that fits to a set of data passed as an array.
    (b) The coefficients of a fitting polynomial of degree n in order of decreasing powers.

(c) The coefficients of a fitting polynomial of degree n in order of increasing powers.

(d) An array of data points which evaluate the fitted polynomial at the points specified by the array x.

37. The subplot command

(a) Allows you to generate several plots contained in the same figure.

(b) Allows you to plot multiple curves on the same graph.

(c) MATLAB does not have a subplot command.

38. Calling subplot(m, n, p)

(a) Divides a figure into an array of plots with m rows and n columns, putting the current plot at pane p.

(b) Plots curve n against m in pane p.

(c) Plots curve p at location (row, column) = (m, n)

(d) MATLAB does not have a subplot command.

39. When writing MATLAB code, to indicate the logical possibility of a NOT EQUAL b in an If statement you write:

(a) a != b

(b) a.NE.b

(c) a <> b

(d) a ~= b

40. The OR operator in MATLAB code is represented by

(a) The .OR. keyword

(b) Typing & between variables

(c) Typing ~ between variables

(d) The "pipe" character |

41. If A is a column vector, we can create the transpose or row vector B by writing:

(a) B = transpose(A)

(b) B = A'

(c) B = t(A)

(d) B = trans(A)

42. Suppose that $x = 7$ and $y = -3$. The statement x ~= y in MATLAB will generate:
    (a) ans = 1
    (b) ans = 0
    (c) ans = –1
    (d) An error
43. To enter the elements 1 –2 3 7 in a column vector, we type:
    (a) [1: –2:3:7]
    (b) [1, –2,3,7]
    (c) [1; –2; 3; 7]
    (d) [1 –2 3 7]
44. To compute the square of a vector *f* of data, write:
    (a) sqr(*f*)
    (b) *f*. ^ 2
    (c) *f* ^ 2
    (d) square(*f*)
45. To generate the fourth order Taylor expansion of a function *f* about the origin, use:
    (a) Taylor(*f*, 4)
    (b) taylor(*f*, 4)
    (c) taylor(*f*), terms('4')
    (d) taylor, *f*, 4
46. The command used to print a string of text *s* to the screen is:
    (a) disp(*s*)
    (b) display(*s*)
    (c) display, *s*
    (d) Disp(*s*)
47. A function is defined as *f* = *a* * *x* ^ 2 – 2. The command solve(*f*, *a*) yields:
    (a) 2/*a*
    (b) 2/*x*
    (c) –2/*x*
    (d) An error: rhs must be specified.

48. The MATLAB ode23 and ode45 solvers:

    (a) Are based on the Runge-Kutta method

    (b) Are based on the Euler method

    (c) Use the method of Lagrange multipliers

    (d) Are relaxation based.

49. A function of time is given by $y = y(t)$ and it satisfies a differential equation. Using ode23 to find the solution generates:

    (a) An array of the form $[t\ y]$.

    (b) The symbolic solution $y = y(t)$.

    (c) An error, MATLAB does not have an ode23 solver.

    (d) An array of the form $[y\ t]$.

50. The average of the elements contained in a vector **v** is obtained by typing:

    (a) Ave(v)

    (b) ave(v)

    (c) mean(v)

    (d) average(v)

51. The pin$v$ command

    (a) Is the proper inverse of a matrix

    (b) Generates the pseudo inverse of a matrix

    (c) Can only be called with symbolic data.

52. Use MATLAB to find

$$\frac{d^{10}}{dt^{10}}(e^{-0.1t}\cos 2t)$$

53. A variable called y has one row and 11 columns. If we type size(y) MATLAB returns:

    (a) 11

    (b) 1 11

    (c) 11 1

    (d) (1, 11)

54. To implement a for loop which ranges over $1 \le x \le 5$ in increments of 0.5 the best command is:
    (a) for i = 1, 5, 0.5
    (b) for i = 1:0.5:5
    (c) for loop i = 1, 5, 0.5
    (d) loop i = 1:0.5:5 do
55. A function $y = e^{-2x}$ is defined numerically for some data range x. It can be entered using:
    (a) y = exp(–x) * exp(–x)
    (b) y = sqr(exp(–x))
    (c) y = exp(–x). * exp(–x)
    (d) y = sqrt(exp(–x))
56. To find the largest value in a column vector **x**, type:
    (a) max(x)
    (b) maximum(x)
    (c) largest(x)
57. The labels used by MATLAB on an x–y plot are specified with:
    (a) The labels command.
    (b) The *x*label and *y*label command.
    (c) The label command
    (d) The text command
58. If we make the call y = polyval(n, x), then polyval returns:
    (a) There is no polyval command.
    (b) A polynomial of degree n evaluated at point x.
    (c) The coefficients of a degree n polynomial.
59. To specify the domain and range used in an x–y plot where $a \le x \le b$, $c \le y \le d$
    (a) axis([a b c d])
    (b) axis(a b c d)
    (c) axis([a, b, c, d])
    (d) axis a, b, c, d

60. To plot a log-log plot, the correct command is:
    (a) plot(x, y, 'log-log')
    (b) loglog(x, y)
    (c) log(x, y)
    (d) logplot(x, y)

In problems 61–70, use MATLAB to calculate the following integrals.

61. $\int \frac{dx}{x\sqrt{1+x^2}}$

62. $\int (\sin x^2)\, 2x\, dx$

63. $\int e^{x^5} x^4 dx$

64. $\int \frac{\cos(\ln(x))}{x} dx$

65. $\int \ln x^2 dx \;\; \int \ln x^2 dx$

66. $\int_0^2 x^3 dx$

67. $\int_0^\infty xe^{-2x}\, dx$

68. $\int_{-\infty}^\infty xe^{-2x}\, dx$

69. $\int_0^b \sinh(x) dx$

70. $\int_{2\pi}^{3\pi} \sin t \sin^2 4t\, dt$

71. Find the area enclosed between the two curves $y = x^2$ and $y = e^{-x}$ for $0 \le x \le 1$.

72. Find the area between $y = x^2$ and $y = -3/2x$ for $0 \le x \le 2$.

73. Use MATLAB to compute

$$\int_0^1 \int_{-1}^1 \int_0^2 x^2 e^{-2xy} z dx dy dz$$

74. Use MATLAB to integrate $f(r, \theta, \phi) = r \sin 2\theta \cos \phi$ over the volume of a sphere of radius 2.

75. When solving a differential equation using MATLAB, the dsolve command expects the user to indicate a second derivative of $y = f(x)$ by typing:
    (a) y''
    (b) diff(y, 2)
    (c) D ^ 2y
    (d) D2y
    (e) D(Dy)
76. To find a solution of

$$\frac{dy}{dt} + 2y = 0$$

The best syntax to use in MATLAB is:
    (a) diff('y' + 2 * y = 0')
    (b) dsolve('Dy + 2 * y')
    (c) dsolve('Dy + 2 * y = 0')
    (d) diff(Dy + 2y)

In problems 77–81, find a symbolic solution of the given differential equation using MATLAB.

77. $\frac{dy}{dt} + 2y = 0$
78. $\frac{dy}{dt} + 2y = -1$
79. $\frac{d^2y}{dx^2} + 4y - 2 = 0$
80. $\frac{d^3y}{dx^3} + y = 0$
81. $\frac{d^2f}{dt^2} + 2f = 4\cos 3t$

In problems 82–90, find the solution to the IVP and BVP indicated. Then plot the solution.

82. $\frac{dy}{dt} - 3y = 0, y(0) = 1.$
83. $\frac{dy}{dt} - 3y = 2, y(0) = -1.$

84. $\dfrac{dy}{dt} + y = \cos t, y(0) = 2.$

85. $\dfrac{d^2y}{dt^2} - 3y + 1 = 0, y(0) = 0, y'(0) = 1.$

86. $\dfrac{d^2y}{dt^2} + y = 0, y(1) = 0, y'(2) = -1.$

87. $\dfrac{d^2y}{dt^2} + y + 4 = 0, y(0) = 1, y'(0) = 1.$

88. $\dfrac{d^2y}{dt^2} + y + 4 = A, y(0) = 1, y'(0) = 1.$

89. $\dfrac{d^2y}{dt^2} + y + 4 = 2e^{-t}, y(0) = 1, y'(0) = 0.$

90. $\dfrac{d^2y}{dt^2} + y + 4 = \sin(5t), y(0) = 0, y'(0) = 1.$

91. Use MATLAB to solve the system of equations:

$$\frac{dx}{dt} = x - 2y$$

$$\frac{dy}{dt} = 2x + y$$

92. Solve the same set of equations if $x(0) = 0$, $y(0) = 1$.

93. Find the inverse of the matrix:

$$A = \begin{pmatrix} 1 & 2 \\ 2 & 1 \end{pmatrix}$$

94. By calculating the matrix inverse of the coefficient matrix, solve the system:

$$\begin{aligned} x + 2y &= 4 \\ 3x - 4y &= 7 \end{aligned}$$

95. Find the rank of the matrix:

$$B = \begin{pmatrix} 1 & -2 & 4 \\ 3 & 2 & -5 \\ 1 & 2 & 8 \end{pmatrix}$$

96. Find the inverse of the matrix in the previous problem.
97. Find the LU decomposition of the matrix B in problem 95.
98. Find the eigenvalues of the matrix B in problem 95.
99. Numerically integrate the function f with data points $f = [1, 2, 5, 11, 8]$; where $0 \le t \le 5$.
100. Numerically integrate $\cos(x^2)$ over $0 \le x \le 1$.

Answers to Quiz and Exam Questions

Chapter 1: The MATLAB Environment

1. 3.9286
2. 2.2×10^3
3. 15.5885
4. False
5. 150.7964 cubic centimeters
6. 1170/1351
7.
```
% File to calculate sin of numbers
x = [pi/4,pi/3,pi/2];
format rat
y = sin(x)
```

Chapter 2: Vectors and Matrices

1.
```
> > x = [0:0.1:1];
>> y = tan(x);
>> plot(x,y),xlabel('x'),ylabel('tan(x)')
```
2.
```
>> z = sin(x);
>> plot(x,y,x,z),xlabel('x'),ylabel('y')
```
3. x = [–pi: 0.2: pi], x = linspace(–pi, pi), x = linspace(–pi, pi, 50)
4. [x, y] = meshgrid(–3:0.1:2, –5:0.1:5). [x, y] = meshgrid(–5:0.2:5)
5.
```
>> t = [0: pi/40:10*pi]
>> plot3(exp(-t).*cos(t), exp(-t).*sin(t),t), grid on
```

Chapter 3: Plotting and Graphics

1. 7.94
2.
```
>> A = [-1+i 7*i 3 -2-2*i];
      >> sqrt(dot(A,A))
      ans =   8.246
```
3. A = [1; 2; 3]; B = [1 2 3] or B = [1, 2, 3]
4. [4; 10 ; 18]
5. eye(5)
6. Array product = $\begin{pmatrix} 16 & 7 & 22 \\ -6 & 30 & -4 \\ 0 & 4 & -16 \end{pmatrix}$, matrix product = $\begin{pmatrix} 31 & 72 & 66 \\ 5 & 34 & 30 \\ -18 & -4 & -8 \end{pmatrix}$
7. Enter the following commands:

```
>> A = [1 2 3; 4 5 6; 7 8 9]

A =

     1     2     3
     4     5     6
     7     8     9

>> B = A([3,3,2],:)

B =

     7     8     9
     7     8     9
     4     5     6
```

8. ans =

```
    -1.2424
     2.3939
     2.8182
```

   det(A) = -198

9. det(A) = -30

```
>> inv(A)*b

ans =

     0.7333
     0.1667
     0.2000
```

10. x =

```
     9.4216
    -0.8039
    -0.9118
```

Chapter 4: Statistics and an Introduction to Programming in MATLAB

1. mean = 156.9677 pounds, median = 155 pounds, standard deviation = 12.2651 pounds.
2. The probability a student weighs 150 pounds is 0.4615.
3. Approximately 0.07.
4. The code is:

```
function cylinder_volume
%ask the user for radius
r = input('Enter Radius')
%get the height
h = input('Enter height')
vol = pi*r^2*h;
disp('Volume is:')
disp(vol)
```

5.
```
x = input('enter x')
      n = input('enter highest power')
      i = 0
      sum = 0;
      while i < = n
        sum = sum + x^i;
        i = i 1;
      end
```

6.
```
for r = 1:n
         sum = sum + 1/x^r
       end
```

Chapter 5: Solving Algebraic Equations and Other Symbolic Tools

1. –14.6882
2. $\frac{-1 \pm \sqrt{22}}{3}$
3. $\frac{-1 \pm \sqrt{5+4\pi}}{2}$
4. $x = 5/2$, ezplot('sqrt(2 * x – 4) –1', [2, 4, 0, 1])
5. The roots are 1.1162, –0.3081 + 1.6102i, and –0.3081 – 1.6102i.
6. $x = 1, y = -3, z = 2$
7. One real root, $x = -0.7035$.
8. –1
9. $x + 1/3 * x \wedge 3 + 2/15 * x \wedge 5 + 17/315 * x \wedge 7 + 62/2835 * x \wedge 9$
10. $1 - 1/8 * x \wedge 2 + 5/192 * x \wedge 4 - 113/23040 * x \wedge 6 + 241/258048 * x \wedge 8$

Chapter 6: Basic Symbolic Calculus and Differential Equations

1. 4
2. The limit does not exist. Calculate the left- and right-hand limits that are –1 and + 1.
3. The asymptotes are at 0, 3.
4. 0
5. (a) $-6/(3 * x \wedge 2 + 1) \wedge 2 * x, 72/(3 * x \wedge 2 + 1) \wedge 3 * x \wedge 2 - 6/(3 * x \wedge 2 + 1) \wedge 2$

   (b) $x = 0$

(c) $f''(0) = -6$

(d) $x = 0$ is a local max since $f''(0) < 0$.

6. diff(y, 2) – 11 * y = –24 * sin(t) + 36 * cos(t) $\neq -4\cos 6t$
7. 4 + exp(–2 * t) * C1
8. –2/3 * exp(–t) + 1/6 * exp(2 * t) + 1/2 – t
9. x = 4 * cos(t), p = –2 * cos(t) – 2 * sin(t) + 2 * exp(–t).
10. x = exp(–t) * (1 + t), p = –2 * exp(–t) – exp(–t) * t + 2

Chapter 7: Numerical Solution of ODE's

1. 0
2. $x_1 = \cos t + \sin t$, $x_2 = -\sin t + \cos t$
3. A plot of the solution is shown in Figure 7-1.
4. 12 points and 41 points.
5. The solution is plotted in Figure 7-2.
6. The solution is shown in Figure 7-3.
7. Runge-Kutta

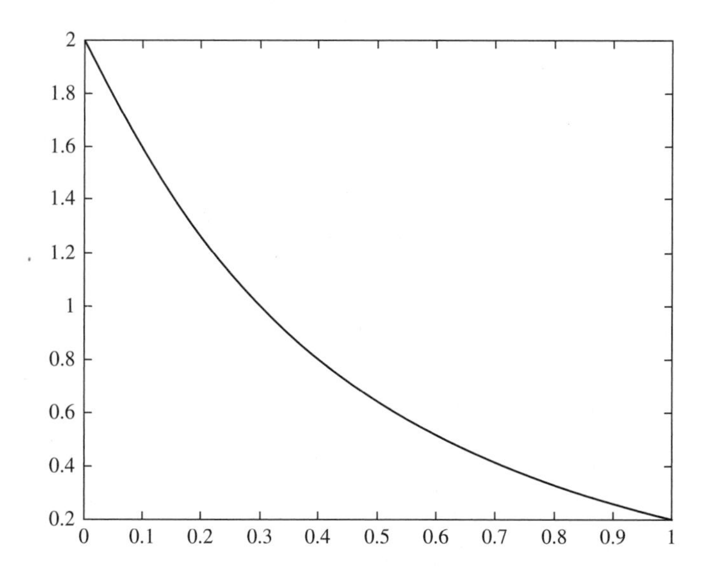

Figure 7-1 Solution to the system in Chapter 7, quiz problem 3

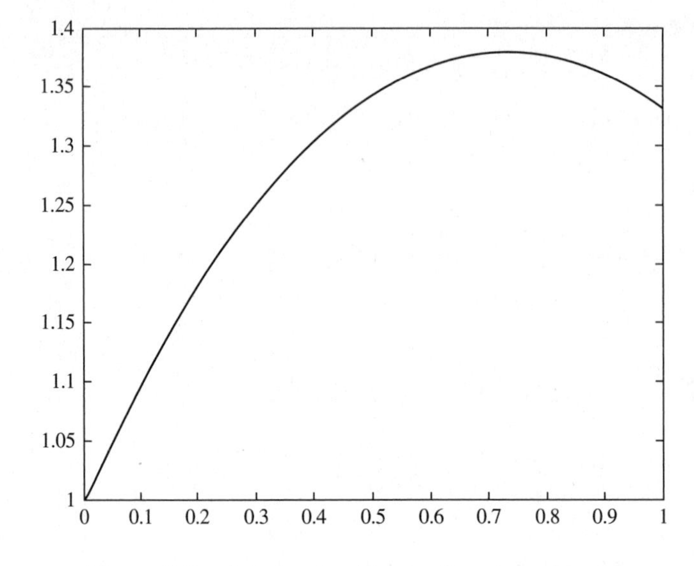

Figure 7-2 Solution for Chapter 7, quiz problem 5

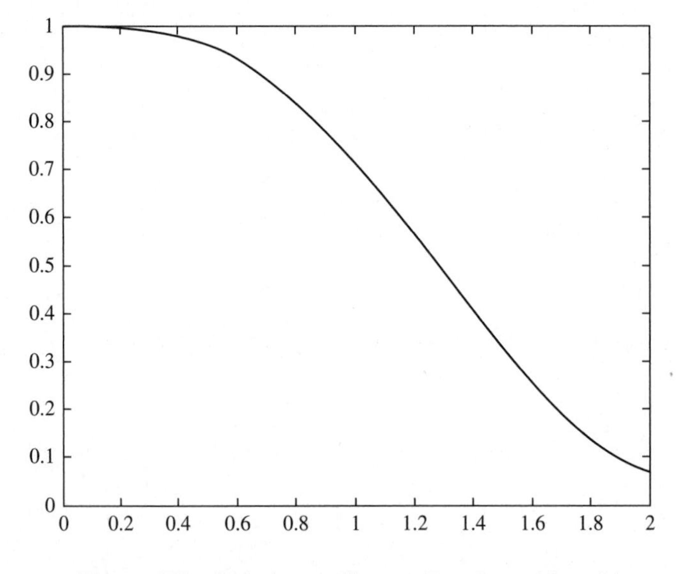

Figure 7-3 Solution to Chapter 7, quiz problem 6

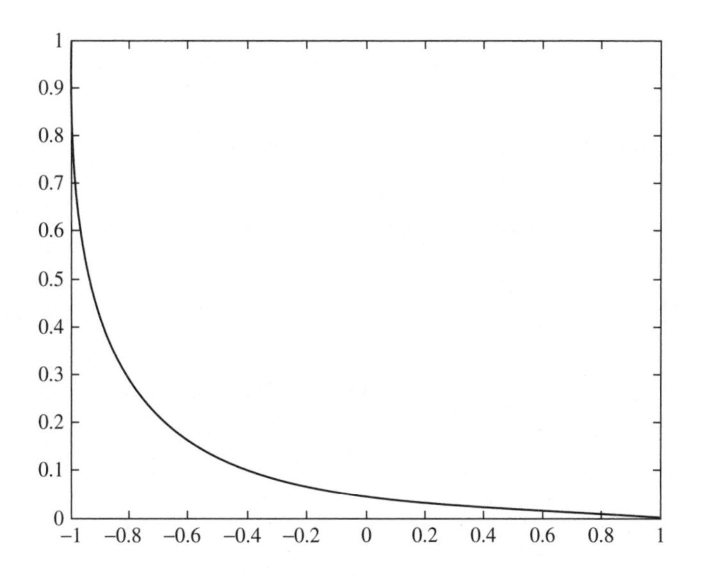

Figure 7-4 Plot of the solution of $\frac{dy}{dt} = \frac{2}{\sqrt{1-t^2}}, \; -1 < t < 1, \; y(0) = 1$

8. Solution is shown in Figure 7-4.
9. The solution is shown in Figure 7-5.
10. The plot is shown in Figure 7-6.

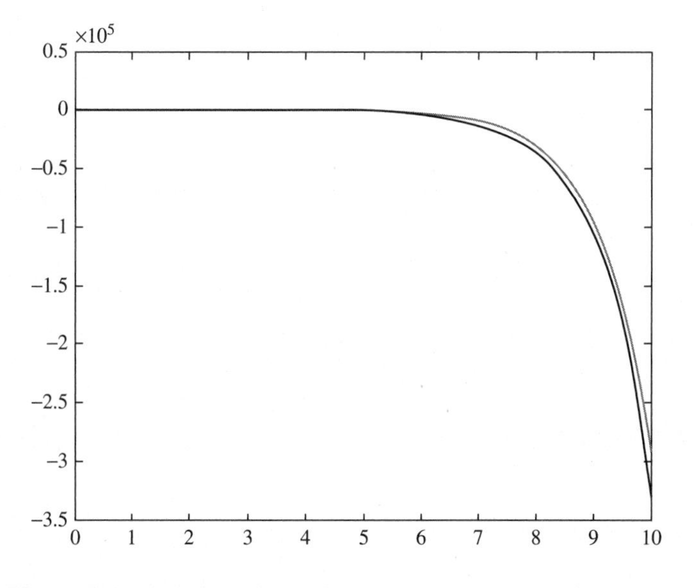

Figure 7-5 Solution of $y'' - 2y' + y = \exp(-t)$, $y(0) = 2$, $y'(0) = 0$

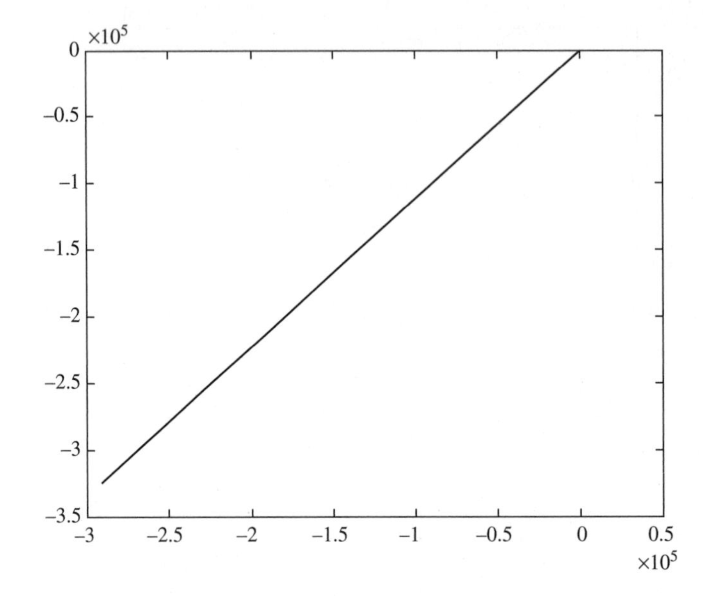

Figure 7-6 Phase portrait for $y'' - 2y' + y = \exp(-t)$, $y(0) = 2$, $y'(0) = 0$

Chapter 8: Integration

1. $1/a \wedge 2 * (-a * x * \exp(-a * x) - \exp(-a * x))$
2. $\log(1/2 * b + x + (x \wedge 2 + b * x + c) \wedge (1/2))$
3. $1/2 * i * x \wedge 2 - x * \log(1 + \exp(2 * i * x))$
   $+ 1/2 * i * \text{polylog}(2, -\exp(2 * i * x))$
4. $1/2 * y - 1/4 * \sin(2 * x * y)/x$
5. 3/10
6. 0.9425
7. $V = \text{int}(\text{int}(\text{int}(r \wedge 2 * \sin(\text{theta}), r, 0, a), \text{theta}, 0, \text{pi}), \text{phi}, 0, 2 * \text{pi})$,
   `>> subs(V, a, 2) = 33.5103 cubic meters.`
8. trapz = –0.1365, relative error 0.34%.
9. The relative error is 0.18629097599734, virtually unchanged.
10. Quadrature integration returns 1.2459.

Chapter 9: Transforms

1. 1/(s + 3) ^ 2
2. 40/(s ^ 2 + 25) – 1/4 * (s + 1) / (1/4 * (s + 1) ^ 2 + 1)
3. 1 – 2 * dirac(t) + 5 * sin(3 * t) (1 is the Heaviside function)
4. cosh(7 * t) – sinh(3 * t)
5. 4 * exp(–1/2 * t) * sinh(1/2 * t)
6. –2 * pi * dirac(2, w)
7. i * pi * (dirac(1, w – 1) + dirac(1, w + 1))
8. exp(–x) * heaviside(x)
9. [7.0000, 3.0000 – 2.0000i, –5.0000, 3.0000 + 2.0000i]
10. 447.7874 at about 40 Hz.

Chapter 10: Curve Fitting

1. $m = 11.8088$, $b = 191.5350$.
2. age_range = [15:1:37];
3. `>> y = m * age_range + b;`
4. 392.2850 pounds
5. 479.3750 pounds
6. 479.3750 pounds
7. $S = 8.3122e + 004$, $A = 1.1184e + 004$.
8. $r2 = 0.8655$.
9. $y = -0.8870x^2 + 58.0577x - 351.6032$
10. 0.9896

Chapter 11: Working with Special Functions

1. 3.5569e + 014
2. 0.494

3.
```
>> x = (1:0.1:2)'; y = gamma(x); A = [x y]

A =

    1.0000    1.0000
    1.1000    0.9514
    1.2000    0.9182
    1.3000    0.8975
    1.4000    0.8873
    1.5000    0.8862
    1.6000    0.8935
    1.7000    0.9086
    1.8000    0.9314
    1.9000    0.9618
    2.0000    1.0000
```
4. 0
5. vol = @ (n, r) r ^ n * pi ^ (n/2)/gamma(1 + n/2), 1.2791e + 003
6. 1.9175e–009
7. The plot is shown in Figure 11-7.
8. 0.2995

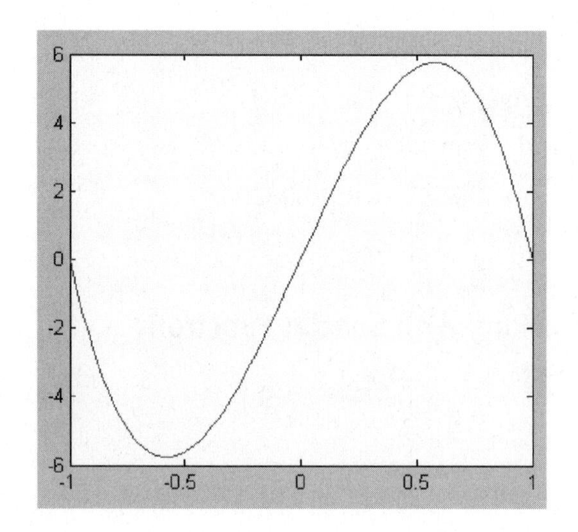

Figure 11-7 A plot of $P_3^2(x)$

9. The table is:

```
A =

         0    0.3550
    0.5000    0.2317
    1.0000    0.1353
    1.5000    0.0717
    2.0000    0.0349
    2.5000    0.0157
    3.0000    0.0066
    3.5000    0.0026
    4.0000    0.0010
    4.5000    0.0003
    5.0000    0.0001
```

10. $\dfrac{1}{3^{2/3}\Gamma(2/3)} = 0.3550$

Final Exam

1. 187.5
2. 2150
3. 16.35
4. 2 * x * cos(pi * x) – x ^ 2 * sin(pi * x) * pi
5. 2/(x ^ 2 – 9) – 4 * x ^ 2/(x ^ 2 – 9) ^ 2
6. 2 * cos(x) – x * sin(x)
7. Use the expand command to get 1 + 4 * x + 6 * x ^ 2 + 4 * x ^ 3 + x ^ 4.
8. $x = 0, x = 1$
9. 1
10. Limit does not exist.
11. infinity
12. 0
13. 0.0107
14. –7.99, –21.11
15. 470547/4 * (x ^ (1/2) + 2) ^ 96/x ^ (3/2) – 14553/4 * (x ^ (1/2) + 2) ^ 97/x ^ 2 + 297/8 * (x ^ (1/2) + 2) ^ 98/x ^ (5/2)

16. solve
17. c
18. b
19. $\frac{5 \pm i\sqrt{11}}{2}$
20. $(x - 4) * (x \wedge 2 + 4 * x + 16)$
21. c
22. 1
23. 9
24. 4.333
25. pretty(f)
26. d
27. a
28. b
29. c
30. b
31. d
32. c
33. c
34. a
35. d
36. b
37. a
38. a
39. d
40. d
41. b
42. a
43. c
44. b
45. b
46. a

47. b
48. a
49. a
50. c
51. b
52. The tenth derivative is:

    –9101406417999/10000000000 * exp(–1/10 * t) * cos(2 * t) + 24836027201/50000000 * exp(–1/10 * t) * sin(2 * t)
53. b
54. b
55. c
56. a
57. b
58. b
59. a
60. b
61. –atanh(1/(1 + x ^ 2) ^ (1/2))
62. –cos(x ^ 2)
63. 1/5 * exp(x ^ 5)
64. sin(log(x))
65. log(x ^ 2) * x – 2 * x
66. 4
67. ¼
68. –inf
69. cosh(b)–1
70. 64/63
71. –2/3 + exp(–1)
72. 17/3
73. 5/16 * exp(–4) + 3/16 * exp(4)
74. 0
75. d
76. c

77. C1 * exp(–2 * t)
78. –1/2 + C1 * exp(–2 * t)
79. sin(2 * t) * C2 + cos(2 * t) * C1 + 1/2
80. C1 * exp(–t) + C2 * exp(1/2 * t) * sin(1/2 * 3 ^ (1/2) * t) + C3 * exp(1/2 * t) * cos(1/2 * 3 ^ (1/2) * t)
81. sin(2 ^ (1/2) * t) * C2 + cos(2 ^ (1/2) * t) * C1 – 4/7 * cos(3 * t)
82. exp(t)
83. –2 + exp(t)
84. 1/2 * cos(t) + 1/2 * sin(t) + 1/2 * exp(–t)
85. exp(3 ^ (1/2) * t) * (–1/6 + 1/6 * 3 ^ (1/2)) + exp(–3 ^ (1/2) * t) * (–1/6 – 1/6 * 3 ^ (1/2)) + 1/3
86. –sin(t) – sin(1) * cos(1)/(sin(1) ^ 2 – 1) * cos(t)
87. sin(t) + 5 * cos(t) – 4
88. sin(t) + (5 – A) * cos(t) – 4 + A
89. 4/5 * sin(t) + 23/5 * cos(t) – 4 * exp(–2 * t) * exp(2 * t) + 2/5 * exp(–2 * t)
90. 4/5 * sin(t) + 23/5 * cos(t) – 4 * exp(–2 * t) * exp(2 * t) + 2/5 * exp(–2 * t)
91. The solutions are:

    x = exp(t) * (C1 * cos(2 * t) – C2 * sin(2 * t))

    y = exp(t) * (C1 * sin(2 * t) + C2 * cos(2 * t))
92. x = –exp(t) * sin(2 * t), y = exp(t) * cos(2 * t)
93. The inverse is:

$$A^{-1} = \begin{pmatrix} -1/3 & 2/3 \\ 2/3 & -1/3 \end{pmatrix}$$

94. $x = 3$, $y = ½$
95. 3
96. The inverse is:

```
 0.2600    0.2400    0.0200
-0.2900    0.0400    0.1700
 0.0400   -0.0400    0.0800
```

97. The LU decomposition is:

```
L =

    0.3333    1.0000         0
    1.0000         0         0
    0.3333   -0.5000    1.0000

U =

    3.0000    2.0000   -5.0000
         0   -2.6667    5.6667
         0         0   12.5000
```

98. The eigenvalues are:

```
1.6177 + 3.2034i
1.6177 - 3.2034i
7.7647
```

99. 31

100. 0.9045

INDEX

A

B

C

D

M

N

O

S

T

U

V

W

X

Y

Z